ASQ's Pocket Guide Series

INTERNAL AUDITING FUNDAMENTALS

Third Edition

Paul Russell

QUALITY PRESS

Milwaukee, Wisconsin

American Society for Quality, Quality Press, Milwaukee 53203

Publisher's Cataloging-in-Publication Data

Names: Russell, Paul B., 1974-, author.
Title: Internal auditing fundamentals , third edition / by Paul Russell.
Series: ASQ's Pocket Guide Series
Description: Includes bibliographical references. | Milwaukee, WI: Quality Press, 2025.
Identifiers: LCCN: 2024951760 | ISBN: 9781636941523 (paperback) | 9781636941530 (PDF) | 9781636941547 (epub)
Subjects: LCSH Auditing, Internal. | Quality control. | BISAC BUSINESS & ECONOMICS / Auditing | BUSINESS & ECONOMICS / Quality Control
Classification: LCC HF5668.25 .R87 2025 | DDC 657/.458--dc23

ASQ advances individual, organizational, and community excellence worldwide through learning, quality improvement, and knowledge exchange.

Bookstores, wholesalers, schools, libraries, businesses, and organizations: Quality Press books are available at quantity discounts for bulk purchases for business, trade, or educational uses. For more information, please contact Quality Press at 800-248-1946 or books@asq.org.

To place orders or browse the selection of all Quality Press titles, visit our website at: http://www.asq.org/quality-press.

29 28 27 26 25 LS 7 6 5 4 3 2 1

QUALITY PRESS

Quality Press
600 N. Plankinton Ave.
Milwaukee, WI 53203-2914
Email: books@asq.org
Excellence Through Quality™

Table of Contents

Chapter 1

Welcome to Quality Auditing

Internal Auditing Fundamentals prepares those new to auditing to conduct internal audits against quality, environmental, safety, and other specified criteria. You may be learning the basic auditing conventions to qualify as an internal auditor or for self-improvement. In either case, both you and your organization will benefit from your new skills. Your organization will benefit because you will be a more effective auditor and you will benefit because you will gain knowledge and learn new skills. Not only will you be learning new skills in auditing, you can also use these skills in other job responsibilities, be able to link requirements to your job, and improve your everyday communication skills by practicing interviewing techniques. After you learn the basics of internal auditing, you may seek more advanced study to qualify as an ASQ Certified Quality Auditor (CQA) or work toward other ASQ certifications. The scope of work for an internal auditor assignment can vary from simple verification of compliance to identification of performance-improvement opportunities. Your organization has objectives that the internal audit program can help achieve.

Even though some people automatically assume the term *audit* refers only to financial matters, that is simply not the case. An audit is some type of formal independent examination of products, services, work processes, departments, or organizations. Conducting an audit is a process, work practice, or service. Some organizations prefer the word *evaluation, survey, review,* or *assessment* instead of the word *audit.* We will use the word *audit* when referencing the process because it is universally accepted and, to experts, it means a certain type of investigation or examination as described in this guidebook.

The audit process steps are shown in Figure 1.1.

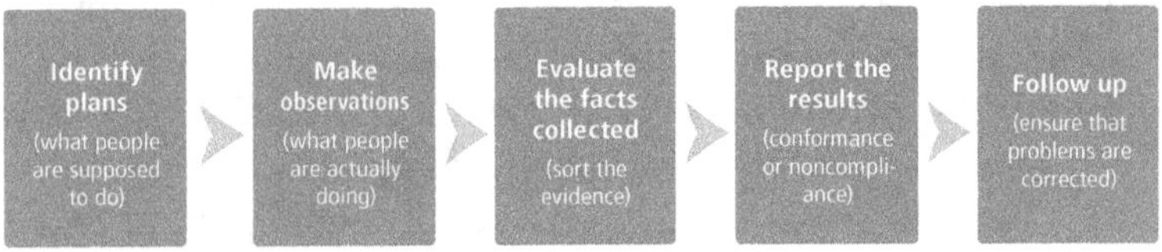

Figure 1.1 The audit process.
J.P. Russell, copyright 2006.

Auditors are entrusted with confidential information no matter the name used for the audit process. Auditors must be ethical in their dealings with the organizations they audit as well as with the general public. People have various feelings about auditors, including fear and respect, but there is also a sense that auditors hold a public trust of honesty and conduct their affairs ethically. When this public trust is broken (e.g., in the Arthur Andersen–Enron case) the public is outraged. Enron was being investigated for possible fraud on a very large scale; a decision was made by their accounting firm Arthur Andersen to shred and destroy the

Enron audit records instead of retaining them or having them available for investigation. At the time of the Enron incident, Arthur Andersen was one of the top five accounting firms in the United States with roughly 85,000 worldwide employees and now, because of the misconduct of a few auditors shredding Enron audit documents, they are out of business. The loss of confidence was quick, and corporations dissolved their ties with Arthur Andersen; the firm itself was indicted for obstruction of justice.

From time to time throughout this guide, we will highlight one of the 20 Basic Audit Principles to emphasize its importance. All 20 audit principles are listed in Appendix M, which the American Society for Quality (ASQ) highlights into four categories (auditor conduct, preparing, performing, and reporting). The first audit principle concerns public trust.

> **Audit Principle**
>
> *Use knowledge and skills for the advancement of public welfare.*

TERMINOLOGY

This chapter is about the terminology of auditing to help you communicate effectively. Your organization may have its names for things that are different from standard audit terms or even different from the dictionary. If the terminology in the text starts to get confusing, consider starting your cross-reference showing the word you are familiar with compared to the more generic terminology. You can start with the examples shown in Table 1.1.

CONTROLS TO EXAMINE

An audit is a process of investigating and examining evidence to determine whether agreed-upon requirements are being met. The definition in the ISO 9000 International Standard is similar: a systematic, independent, and documented process for obtaining objective evidence and evaluating it objectively to determine the extent to which the audit criteria are fulfilled (clause 3.13.1 in the ISO 9000:2015 standard). An effective audit depends on how information is gathered, analyzed, and reported. The results may verify conformance or indicate noncompliance with

Table 1.1 Example terminology cross-reference table.

No.	Universal terminology	Your organization's term
1	Audit	Assessment, evaluation, check-up
2	Survey	Review
3	Audit program department	Regulatory compliance department
4	Employee	Associate

Table 1.1, continued

No.	Universal terminology	Your organization's term
5	Customer	Client, patient, member, passengers, students
6	Client	Program manager, quality/safety/environmental manager
7	Audit program manager	Compliance director

rules, standards, or regulations. A quality audit is linked to quality requirements, environmental audits to environmental requirements, financial audits to financial statements, and safety audits to safety rules and regulations. One of the things that makes an audit different from an inspection is that individuals performing an audit must be able to do so impartially and objectively. This means that the person performing the audit must be independent of or have no vested interest in the area being audited. The level of independence necessary to ensure impartiality and objectivity will vary by industry, type of organization, risks involved, and organizational culture. Sometimes independence from the area audited must be formally verified.

INTERNAL AND EXTERNAL AUDITS

All audits are either internal audits or external audits. Figure 1.2 shows how audits are classified as first (internal), second (external), and third (external) party.

Think of your organization as the circle in the figure. Internal or first-party audits are conducted inside the circle. You must go outside the circle to conduct external or second-party audits (audit your suppliers).

On the right-hand side of the figure is an area designated for third-party audits. Third-party audits are independent of the customer-supplier relationship. Third-party audits may result in certification, license, or approval of a product, process, or system by an independent organization. Your organization may have its quality system or environmental system registered by a third-party certified body, notified body, or registrar,

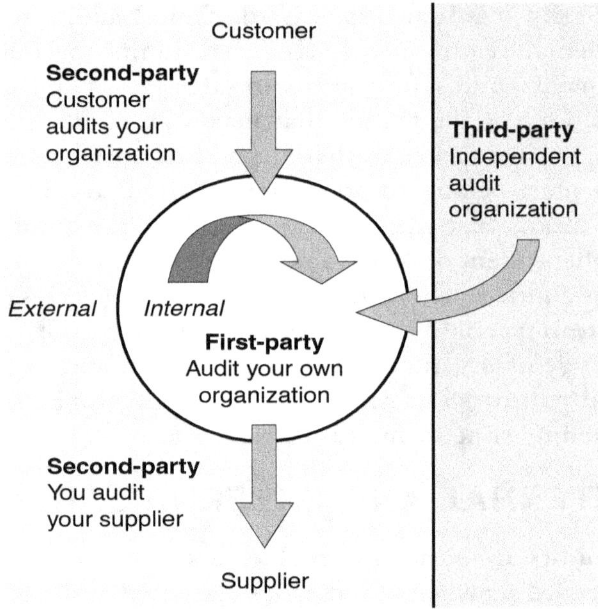

Figure 1.2 Audit classifications.

or licensed by a government oversight agency. While this book was specifically written to address internal auditing, you may become aware that the overall process (from a high level) is followed for any type of audit. One of the reasons internal audits are conducted is to help prepare organizations for audits conducted by external audit organizations (for example, customers, certified or notified bodies, registrars, and government agencies).

AUDIT TYPES

Audits are also classified by area (process, system) or object (product, service) of the audit. You may be assigned to conduct a system, process, or product audit. Different audits may require different methods, personnel, or equipment.

The *product audit* (or can also be a service audit for an organization in the service industry), the smallest circle in Figure 1.3, determines if tangible characteristics and attributes of a thing are being met. Typically, an auditor checks the object or service to ensure that it has the proper markings, weight, size, viscosity, smoothness, amount, hardness, color, texture, placement, arrangement, count, and so on.

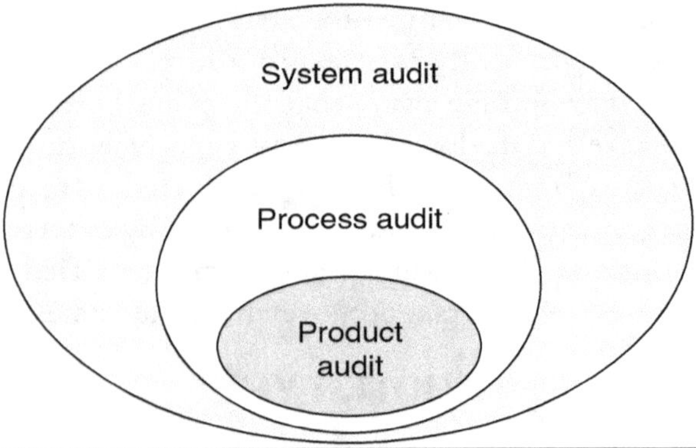

Figure 1.3 Different types of audits.

The auditor checks the object or service against a predetermined set of characteristics or attributes. A product audit is just like an inspection (bicycle, hotel room, final presentation slides, a football helmet, etc.) except there must be some level of independence and the audit results are not used to approve the release of a product or delivery of a service. In fact, a product audit can be done after an inspection takes place (that is, after the product has been approved to be released).

A *process audit* determines whether process requirements are being met. During a process audit, the auditor will examine an activity or sequence of activities to verify that inputs, actions, and outputs are in accordance with an established procedure, plan, or method. Outputs can be compared to objectives to determine effectiveness and efficiency. A process audit may examine a particular task such as stamping, welding, serving, sterilizing, filing, cleaning, transacting, or mixing, or sets of processes within processes such as manufacturing, delivering, purchasing, or designing. The activity examined during a process audit normally is described with a verb, indicating that an action is taking place. A process audit normally follows a process from beginning to end or end to beginning. Do not underestimate the power of the process audit. The whole system of a business is made up of many processes working together to achieve business objectives.

A *system audit* determines whether system requirements (manual, policy, standards, regulations) are being met. When processes are interrelated and interacting, you have a system. Sometimes it is easiest to think of the system as the entirety of an organization or business. A system is made up of processes organized to achieve an objective such as quality, safety, or income. During a system audit, you may examine the operation of a department, company, division, or program. Auditors may conduct a product or process audit as part of a system audit. Typically, an auditor will audit an organization against clauses of a quality, safety, or environmental management system standard.

It may help you to think of this type of audit classification as zooming in or out of a picture. For example, in the picture of the racers (Figure 1.4):

- A *product audit* would be checking the helmet or helmets for such attributes as size, color, hardness, markings, identification, webbing, chin strap adjustment, and so on, against requirements (specifications). You may decide to check the team helmets, check all the helmets at the skating rink, or visit the manufacturer

Figure 1.4 Example-type of audit classification.

and sample several helmets. You can do the same thing for a service such as inspecting for the proper arrangement of a cleaned hotel room, cleanliness of a rental car, proper storage of gear before a flight, and so on.

- A *process audit* may evaluate the methods used for skating during a race or methods for skating in a sharp turn. You may ask about training, techniques to be employed, type of equipment required, measures for determining a successful turn, adjustments for ice conditions, and equipment prep and maintenance.

- A *system audit* may evaluate the management of the skating team or management of the skating arena. You may be interested in how events are scheduled, communication with team members, how changes are implemented, preventive maintenance programs, operating the box office, maintaining and operating the Zamboni, how customer needs are determined, and so on.

Most internal audits are either process or system audits. Many organizations divide up their system into little pieces or elements and assign each of their internal auditors to one. Other organizations may divide up the system into big chunks and assign teams of auditors to evaluate them.

KEEN OBSERVATIONS

Regardless of the type of audit, an auditor must be good at observing and reporting factual information.

The person conducting the audit is the *auditor.* Other equivalent descriptive words are *evaluator, assessor, examiner, reviewer,* and so on. The organization being audited is called the *auditee.* Any type of organization can be an auditee (your department, a corporation, a government agency, a nonprofit organization, a retail sales store, a manufacturer, and so on). The person or organization who requested the audit is the *client.* Audits are only conducted when someone or some group requests one. You might think of the client as the person who has the authority to assign you to do an audit. This person is one of the customers of the audit service, to whom you are accountable. This person (the client) normally is your boss, the audit program manager, or the quality/environmental/safety manager. For example, in an internal audit setting, while your final audit report is for the client (your manager), it will usually then be approved for sending it on to the auditee (when the auditee is not the client).

The next several chapters will take you from getting the audit assignment and reporting findings to ending the audit by completing follow-up actions.

Chapter 2

Getting the Assignment

The first phase of the audit is getting agreement among interested parties and specifying the job assignment: finding out who, what, when, where, and why (see Figure 2.1). Normally the person responsible for the audit program or the lead auditor will contact you about conducting the audit. This person could be the audit program manager, quality manager, compliance director, safety supervisor, management representative, director of environmental affairs, and so on. The person who has the authority to require the audit is called the *client*. The client could be one of the people mentioned or someone entirely different, such as the VP of operations.

It is very important to fully understand the assignment because you as the auditor will have some decisions to make. You have been contacted because the audit program manager decided that you are qualified to conduct the audit. If you do not think you are qualified or if there is a possible conflict of interest, you need to tell the audit program manager or lead auditor immediately.

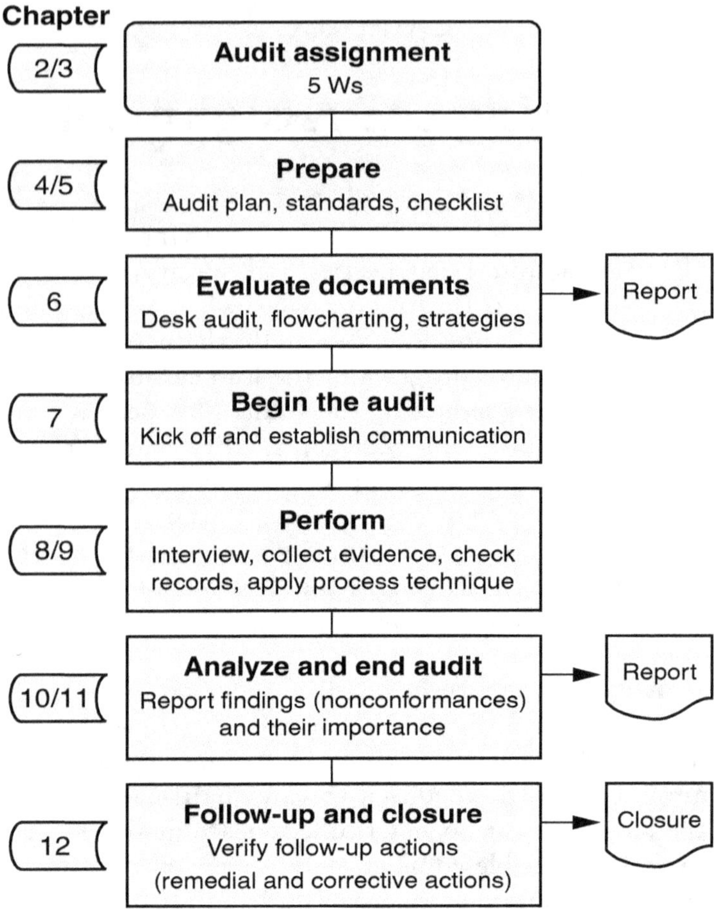

Figure 2.1 Auditing process steps.

ACCEPTING THE ASSIGNMENT

You should be told the area to be audited, the standard or procedure to audit against, and the date and time or time frame. Ask yourself three questions:

Question 1: Are you available for the audit? Yes or No

Availability may include the means, budget, and permission. Do you have a schedule conflict? Are there any financial constraints such as budget or spending limitations? Do you have enough security clearance for the audit? Are you working on another project that has a higher priority? If you are not available on the dates requested, you may provide alternate dates for consideration.

Question 2: Are you free of any conflict of interest? Yes or No

For internal company audits it is impossible to be totally independent. Based on the situation, you will need to declare any potential conflict of interest. For internal audits, acceptance of gifts as a cause for a conflict of interest is unlikely.

Employee relationships and auditing your own work are the two major areas that could result in a conflict of interest. If you were working in the department six months prior, then there is a very good chance you could be auditing your own work during the audit.

Audit Principle

Use knowledge and skills for the advancement of public welfare.

Examples of conflicts of interest are:

1. You are being asked to audit something you developed.
2. A close friend or relative works in the area.
3. You are currently doing other work for the department or area being audited.
4. There is bad blood or personality conflict with personnel in the area to be audited.
5. There has been acceptance of or promise of a gift having value.
6. You are a previous employee of the department or area to be audited. (Note: Some audit programs require a waiting period before auditors can audit prior work areas.)
7. You have a previous close working relationship with the people in the area to be audited.

Internal audits by their very nature may make it impossible to avoid all conflicts of interest. During internal audits you should be on your guard for any biases that could cloud your judgment. The goal is to ensure that the integrity of the audit service is maintained.

Also, some audit program situations are more formal than others, depending on the organization's needs. For example, you may be a full-time compliance auditor who works for the regulatory compliance director who reports directly to the president. In some cases, independence from the area to be audited is not only desirable, but it may also be a requirement.

In other situations, auditors may only be part-time and normally have other full-time duties. For example, you may work in the distribution, quality control, or purchasing department and only conduct one audit each quarter of the year. A potential conflict of interest may be more likely to occur when part-time auditors are used. What is important to remember is: *the goal is to ensure that audits are conducted in an objective and impartial manner.*

Organizational culture plays a major role in determining the amount of independence needed to assure objective and impartial audits. In some organizations, relationship issues are not a concern because everyone is expected to be open, honest, and willing to change as part of their team contribution.

Conflicts of interest may shed doubt on the objectivity and impartiality of audit results. This will adversely affect the integrity of the entire audit program.

Question 3: Do you feel you can do a competent job? Yes or No

Do you feel comfortable auditing your assigned area against the standard selected? If you have been trained and qualified by your organization, you should be able to do the job. Perhaps you were assigned to the Computer Information Systems Technology Solutions Group (CISTSG), however, and you are still trying to figure out document scanners, USB connections, and iPhones. Or you may be missing a certification or a government clearance rating. If so, let the lead auditor or audit program manager know. For a system audit, you need to understand the requirements of the system standard being audited against and have a general understanding of how the area you will be auditing operates. For example: you might need to interpret the requirements of ISO 13485 and their application. For a process audit, you will need to understand the procedures and the process being audited. For example: if you were going to audit a coating process, you should be familiar with good coating practices.

> **Audit Principle**
>
> *Assigned auditors must be competent/qualified.*

If you can say yes to all three questions, accept the assignment with enthusiasm.

Auditors should be willing to sign and attest to their willingness to abide by a code of conduct. An example Auditor Code of Conduct has been provided for you to sign in Appendix J. We recommend you review the

Auditor Code of Conduct and sign it as other auditors must do.

Next you will learn the additional information you will need before you start performing work in preparation for the audit.

Chapter 3

Audit Process Inputs (Purpose and Scope)

You will need certain basic inputs before you can plan for an upcoming audit. Key inputs include:

A. When and where is the audit scheduled?

B. What area(s) are to be audited (for example, department, group, area, or process)? This is called the *scope.*

C. What standards are you auditing against (for example, ISO 9001, IAF 16949, 21 CFR 210 and 820, FAA Title 14 Chapter III, ISO 14001, OHSAS 18001, ISO 13485, ISO 22000, operations manual, work instructions)?

D. What is the purpose? Why do the audit? Is it to verify compliance or conformance? Prepare for government or external audit organizations? Verify contract requirements? Train new auditor candidates? Verify implementation of a new process? Identify opportunities for improvement? Verify corrective actions from prior audits. and so on. What type of audit is it—System? Process? Product?

E. Do any other audit services need to be performed with this requested audit (for example, desk audit, closeout of prior nonconformances, product audit)?

WHEN AND WHERE IS THE AUDIT SCHEDULED? [WHEN]

You will need to know the time and place of the audit so that you can make the necessary arrangements. Getting to the audit site can range from walking down the hall to flying to operations in or out of the country. This book covers the basic process of the internal quality audit with the assumption that it will be in person. Perhaps a portion, or all, of your audit will be performed remotely/virtually. If this is the case, please view additional resources from experts such as Shauna Wilson on remote audit plans, techniques, and considerations. Once you understand the fundamentals here, it will ease the understanding of retaining principles while learning the nuances of performing remote audits later.

WHAT AREA(S) ARE TO BE AUDITED? [WHAT AND WHERE]

Will you be auditing administrative areas such as records control, or technical areas such as research, or operation areas such as production, loading, or treating? The scope of the audit may encompass location, product line, market, customer, function, department, branch, and so on. How much of the organization will be looked at and how many departments will be involved? If it is a multiple-shift operation, all shifts may need to be audited, possibly involving evenings, nights, and weekends. All auditors need to know the parameters of the audit investigation.

Once the audit starts, the scope should not be changed. Only the audit boss (client) can change the scope of an audit once it is agreed upon. If the scope is changed, auditors should be given sufficient time to prepare.

WHAT STANDARDS ARE YOU AUDITING AGAINST? [WHAT]

You need to know which standard(s) and which elements of the standard you are being asked to audit against. Auditors do not make up the rules; auditors audit against existing rules, requirements, procedures, instructions, and so on.

The requirements can be found in documents and other documented information. When utilizing documents, it is popular to think of documents as coming from different levels (see Figure 3.1).

You will be told which standards to audit against. However, your assignment could be very general and only state, "Audit against standard XYZ and the company's management system documents." This leaves you with the responsibility of specifying the applicable quality, safety, and environmental system documents and clauses of the standard that apply. It is convenient to think of documents as having higher-level requirements flowing down to lower-level requirements. As an auditor, you are also responsible for understanding the requirements of the standards and documents being audited against. If you are not familiar with the standard(s) or management system documents, it will be necessary to take a training class or to initiate a self-study program.

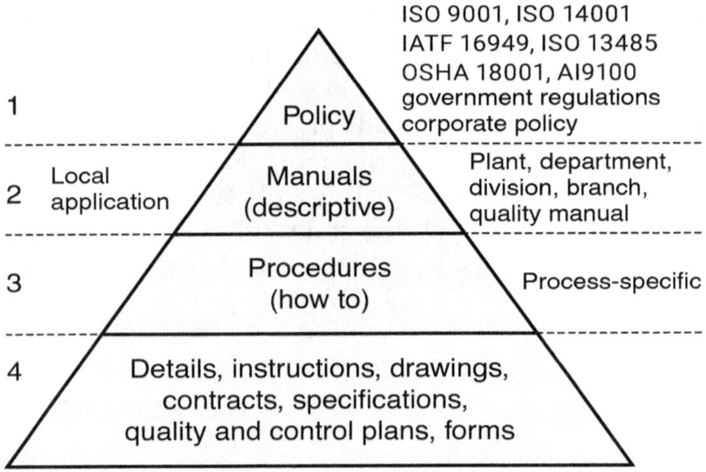

Figure 3.1 Document levels.

When possible, at least two document levels (see Figure 3.1) should be audited. For example, auditing against requirements in both ISO 9001 and procedures. For a process audit, you might use a procedure and work instructions or work instructions and product/service specifications.

Some audits use an entire standard and some audits use only a portion of a standard. If you have been assigned as lead auditor of an audit team, you may be given the standards to audit against by the client. Then it will be up to you (the lead auditor) to make individual audit team member assignments (Oscar gets customer satisfaction in clause 9, and Rachel gets training from clause 7, etc.).

WHAT IS THE PURPOSE (OBJECTIVE) OF THE AUDIT? [WHY]

The purpose (objective) is the why of the audit. By definition an audit is conducted to determine the extent to which agreed-upon criteria have been met. In regulated industries and organizations that have registered quality/environmental systems or licensed processes, audits are used to establish conformance or nonconformance to standards. For example, conformity may result in registration of the quality/environmental system, supplier approval, or product license; nonconformity may result in suspension of registration, supplier disapproval, or license suspension.

When you get the assignment, you should also be told the purpose (objective) of the audit.

Following are some example purpose statements for internal audits:

- To determine the finishing area's adherence to ISO 14001 and EMS procedures
- To verify that X product is being processed in accordance with contract XYZ and cGMPs
- To determine conformance to ISO 9001 for purposes of preparing the area for an external compliance audit (registrar, government agency, certification body)

In general, the purpose of a system or process audit is to determine, confirm, or verify compliance to the audit criteria.

NEED FOR OTHER AUDIT SERVICES

Other internal audit services may be requested and may be included in the purpose statement. Other purposes can include:

- Verifying that corrective actions from prior audits have been implemented
- Assessing progress toward implementation of a quality/environmental system
    - Assessing the progress of project implementation at one or all locations
    - Identifying areas for improvement
    - Evaluating capacity to ensure compliance
    - Evaluating effectiveness in meeting management objectives
    - Assessing process validation status
    - Preparing for a customer audit
    - Assessing on-site supplier services (for example, observing calibration checks or equipment maintenance)
    - Training new auditors

Be sure to plan your time according to the work required. Key questions and concerns should be resolved by the lead auditor or audit boss before the audit.

The next chapter is about preparing for the upcoming audit. The better prepared you are, the more effective the audit will be.

Chapter 4

Preparing for the Audit

You are aware of the upcoming scheduled audit and need to start thinking about what you must do to prepare. Preparation includes: (1) selecting the audit team member(s), (2) preparing an audit plan, (3) understanding audit objectives, (4) identifying requirements, (5) preparing or securing a checklist, and (6) determining the data collection plans. Preparation steps will also be discussed in Chapter 5.

THE AUDIT TEAM

The audit team may be one person or a team of two or more. The lead auditor or audit team leader and audit program manager are responsible for ensuring that there are sufficient resources (such as, auditors) to accomplish the purpose for the defined scope. If the purpose, scope, and resources don't match up, one or more of them must be changed (for example, add more auditors, reduce the scope, or change the purpose). The number of auditors selected must be adequate to carry out the audit in the time allocated. Some organizations publish guidelines for determining the audit time needed for a certain purpose and scope. If the guidelines require two audit days, two auditors should be able to complete the audit in one day. If no guidelines exist, the lead auditor or audit program manager may estimate audit days. Some industries may require a certain amount of time allotted for audits due to current risk-based awareness or perhaps past uncovered issues. The availability of the auditors, schedule conflicts from the auditee, and many other considerations must be factored in to come up with the number of audit days on site.

(See Appendix F: Audit Time Considerations.)

Audit Principle

Ensure that sufficient resources are available to accomplish the purpose of the audit.

Audit team members are responsible for gathering audit evidence of conformance and non-conformance of the area audited. Audit team members analyze data and report nonconformances to the lead auditor. Audit team members report to the lead auditor.

Every audit has a lead auditor, even if there is only one person conducting the audit. The lead auditor is responsible for preparing the audit plan, conducting opening and closing meetings, analyzing all findings to be reported, and reviewing and submitting the final report. The lead auditor is responsible for the performance of the audit team and for initiating and maintaining communication with the audit program manager and auditee organization (unit). The lead auditor normally reports to the audit program manager for matters concerning the audit. There is a certification for a "Lead Auditor" in some fields where a person demonstrates certain abilities to a body granting the specific certification to the person; however, that sort of certification is not relevant to what is being discussed in this book (that is to say, a certification with a title is not the same thing as actually being the lead auditor of an audit).

CONTACT THE AUDITEE AND ISSUE AN AUDIT PLAN

As the audit date approaches, you (as the lead auditor) will need to contact the auditee. It is important to make contact to confirm the upcoming audit. This will avoid any miscommunications about the time of the audit and what is going to be audited. You should always follow your organization's guidelines on when and how you contact the auditee. Some organizations may require contact a month in advance and others may require only two weeks.

Some organizations conduct surprise or unannounced internal audits. This may be a management preference or company policy as practice for unannounced regulatory agency audits. Surprise audits normally are compliance audits that emulate the same methods used by regulatory agency auditors. Surprise compliance audits help ensure that the organization is in a state of readiness for external oversight and are normally limited to that objective. However, whether two weeks or two minutes ahead of time, the lead auditor is responsible for contacting the auditee to confirm the audit schedule.

The lead auditor has the responsibility for making the final arrangements. If you are on a one-person team, you are automatically the lead auditor.

When you make contact, keep in mind the auditee may not totally understand the audit lingo. Ensure you both understand when you go over the following audit information:

- Purpose
- Scope
- Standards and procedures that will be audited against
- Audit team members
- Overall audit schedule for the area (start and end times)
- When the auditee can expect the final report

The schedule should be mutually agreed upon so that there will be no surprises. Never just show up and start an audit unless conducting a surprise audit is a mutually agreed upon audit strategy (in which case the bulleted information above was already agreed upon and the dates/times will be the only surprise). There are some situations in which management may request a surprise audit (for example, to uncover wrongdoing) but then it is not a normal audit. Because of their nature (we don't trust you), surprise audits can create a "we (the auditee) versus them (the auditors)" mentality if not properly managed. Remember, you all work for the same company, so you are all on the same team. Normally the auditee is notified in advance of the planned internal audit. Figure 4.1 shows the flow of information to and from the auditee.

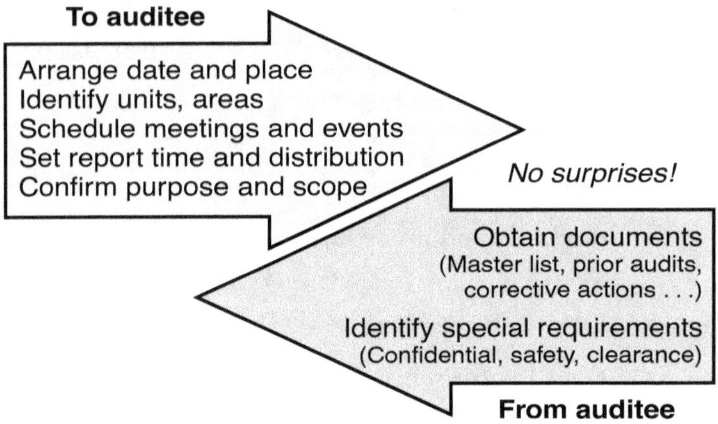

Figure 4.1 Auditor-auditee communication flow.

Before the audit (at initial contact or later) you should obtain any needed documents and records or determine their location. Be aware that many documents and records may need to be safeguarded. Some information is sensitive and may have restrictions for legal, competitive, or security reasons. Internally, you may be able to obtain more documented information actually in your hands to review ahead of time than if you were an external auditor, so safeguarding and being familiar with company policy is a must.

> **Audit Principle**
> *Protect auditee property entrusted to you.*

For internal audits, it is perfectly okay to ask the auditee representative if there is something in particular they want the auditors to examine within the scope of the audit. This could be a new process, a change since the last audit, an historic problem area, or a source of recent complaints. The scope is not being changed, but the auditee's specific needs may be a more prominent factor in your interview and sampling plans. If you think additional audit time is needed, discuss your needs with the audit program manager or client. Follow up your contact with the auditee by issuing the information as an audit plan, sending a copy of the work order (if there is one), or including the information in a message to the auditee (memo, text message, notification letter, e-mail). The amount of formality depends on your organization's situation and culture.

> **Audit Principle**
> *Communicate agreed-upon information to the auditee such as audit times, purpose, areas to be audited, and standards to be audited against.*

If you want to be formal, send out a notification letter along with the audit plan. According to our formal audit rules, the notification letter should be signed by the client. There should be an audit plan for every audit. It may be thought of as your contract with the auditee. It spells out the parameters for the auditing service. See Appendix A: Example Audit Plan, Appendix B: Example Work Order, and Appendix G: Example Notification Letter. A formal notification letter is normally used for second-party and third-party audits. First-party audits are usually less formal and communicated in the form of an email or, in extreme cases, as informally as a quick phone call.

MAKE A LIST

Before you start auditing, list the information, documents, records, standards, and so on that you will need.

Up to this point, there has been a lot of planning and not much action, but good planning and preparation are the keys to an effective audit. Next is the discussion of audit techniques needed for the investigation.

Chapter 5

Identifying Requirements and Planning

No one is born with the knowledge and skill to conduct a proper audit. Auditing is the application of various techniques to collect factual evidence relative to the standard being audited against. Auditing can be hard work, but if you are successful in confirming conformity or identifying opportunities for improvement, you will feel a great sense of accomplishment.

AUDITING OBJECTIVES

Looking at the definition of an audit in ASQ/ANSI/ISO 19011 (*Guidelines for auditing management systems*) and ASQ/ANSI/ISO 9001 clause 9.2 (Internal Audit), we can identify two primary audit objectives for determining conformance:

- Determine the extent of conformity or compliance of the system or process to the defined audit criteria. The audit criteria may be internal documents such as procedures or work instructions or external documents such as ISO 9001, IATF 16949, ISO 13485, IA9100, GMPs, OSHA, FAA, corporate policy, and so on.

- Determine whether the management system controls are effectively implemented and maintained: *implemented and maintained* means the controls are deployed and people are following the rules (procedures, methods, manuals). Is there ongoing adherence to the rules?

Beyond these two basic auditing objectives, other objectives could include: determining the *capability* to meet requirements, evaluating the *effectiveness* of the management system, assessing validation status, and identifying potential *improvement* areas.

When there are external requirements (such as in ISO 9001, GMPs, EPA regulations), you should check to see whether the auditee has addressed the external requirements in some manner. Every place where there are required actions or promises in organization procedures, work instructions, or other methods, you can check to determine whether they have been implemented and maintained. This auditing technique is called the *requirements technique*. Every place where there is a requirement for a tangible deliverable such as a schedule, record, procedure, flowchart, or log, you can check to ensure that it exists. When procedures are required, you can verify that they exist and have been implemented and maintained. This technique is very efficient and traceable to each requirement. This is also called the *element technique* because elements (that is, clauses) of a standard are selected and used as the audit criteria. For example, if a department is required to validate its processes, you could audit them against

ISO 9001 clause 8.5.1f on validation of processes. If the laboratory calibrates equipment, you could audit them against applicable calibration requirements clauses like 7.1.5 in ISO 9001.

In Chapter 9 we will discuss using process auditing techniques. The *process technique* is to follow a process from beginning to end instead of following the elements of a standard. In a process audit, the process may be followed in real time, observing the technician or operator each step of the way. For system audits, an auditor may use an order number or project number to trace an order from acceptance to delivery. Process or PDCA techniques are also very handy when requirements are vague ("do your best to ensure that the kitchen is clean" versus "run the dishwasher and sweep the floor every day"), or the effectiveness of the requirements technique starts to break down because auditors and auditees are unsure of what the requirements mean ("do your best to ensure that it is clean").

However, most standards have prescriptive requirements that organizations can be audited against. Auditors should be prepared to employ several techniques during the investigation to verify conformance to agreed-upon audit criteria.

THE REQUIREMENTS

Requirements come from many different sources. Your organization adheres to mandatory regulatory requirements, customer-imposed requirements, contractual requirements, and self-imposed requirements. In most internal audit programs, someone has already decided which requirements you should audit against and may provide a ready-made checklist (list of requirements) for you to use. However, you need to be able to recognize a requirement (know it when you see it) because all audited requirements must be traceable to a source. It is good to remember that an internal auditor is not the police, but someone who collects data to show the requirements are being met (most requirements are usually being met). The expectation is that the requirements are all being met, and it is the job of the auditor to verify that with the objective evidence collected. If the auditor cannot collect enough evidence showing the requirement is being met for something, then the auditor will need to have data showing the requirement was not being met.

Many contracts, codes, and standards use auxiliary verbs to identify a requirement as well as an expected degree of compliance. Some auxiliary verbs may denote mandatory compliance, while others are used to denote suggestions, optional requirements, or guidance. These auxiliary verbs are used as indicators of the importance of certain requirements. See Figure 5.1.

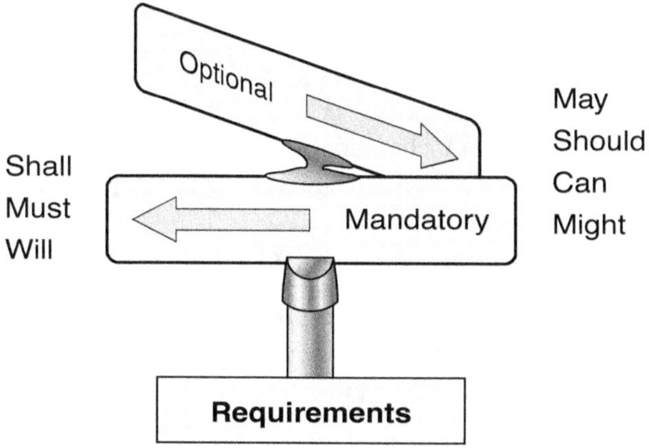

Figure 5.1 Requirements auxiliary verbs.

Mandatory Requirements

Shall. The organization *shall* conduct internal audits at planned intervals (ASQ/ANSI/ISO 9001, clause 9.2.1, Internal Audit). Obsolete and outdated labels, labeling, and other packaging materials *shall* be destroyed (GMPs, CFR 21 211.122 (e), Materials examination usage criteria). The organization *shall* document procedures for the validation of processes for sterilization (ISO 13485, clause 7.5.7, Particular requirements for sterile medical devices).

Must. Technical requirements of the following nature *must* be included by statement or reference (Mil-Q-9858A, clause 5.2, Purchasing Data).

Optional Requirements

Should. Objectives *should* be established for an audit program, to direct the planning and conduct of audits (ASQ/ANSI/ISO 19011, clause 5.2.1, Objectives of an audit program).

May. This examination *may* include the provider's written information (for example, catalogs, leaflets) and evaluation reports (ASQ/ANSI/ISO 10015, clause 4.3.5, Selecting a training provider).

Can. The operations and activities of an organization *can* have a variety of characteristics. Examples of characteristics include wastewater discharge, energy use, air emissions, waste, temperature, and so on (ASQ/ANSI/ISO 14001, clause 9.1, Monitoring and measurement).

However, there is no guarantee that the standards you are going to audit against follow the above conventions and there is no requirement that they do so. When you read a standard or procedure, you should be aware of the authoring conventions being used. Internal organization procedures may not follow any set convention. When there is no established convention, auditors should look for action verbs, such as 'to do' statements, to identify promised or required actions.

Requirements are found in different documents issued at different levels within and external to the organization. It is popular to depict the documents as a pyramid (see Figure 3.1) with external requirements being at the top and internal detailed instructions at the bottom. Some document diagrams depict records as documents. Some documents are made to be written on to become records when it is finished. Dictionary definitions support that a record is a type of document, but many professionals find it less confusing if documents and records are considered two different things. Many consider a document as something that happens before an activity or an intention (plan, procedure, specification, instructions, form, or checklist) and a record as something that happens after an activity or a fact of what occurred (that is, the results of an activity entered on a form or spreadsheet).

An auditor can audit against requirements in external standards (such as ISO 9001, ISO 14001, OHSAS) or internal self-imposed controls. What an auditor should never do is to make up the rules they think the auditee should comply with.

Audit Principle

Verify conformance to agreed-upon requirements (the rules).

CHECKLISTS

A checklist is a 'must' auditor tool that is used to match what the auditee is supposed to be doing with what is being done. A checklist is like a grocery list. You put down the things you are going to check for and you prepare the list before you go to the store. It also provides a place to put your notes, keep track of your interviews, and record observations (audit evidence). The checklist should be designed to help you, the auditor, during the performance phase of the audit. A checklist may contain questions or statements, but all should be linked to a requirement. An auditee has every right to ask for the source of any requirement they are being audited against. You should be able to respond chapter and verse with the standard, procedure, clause, paragraph, and so on.

You may be provided a canned checklist, but you must still know how checklists are constructed and how to add checklist questions that need to be answered by the auditee.

The purpose of a checklist is to aid the auditor in the gathering of information. It helps guide the investigation and provides a place to record information. A checklist can be a list of questions, a series of statements, or even keywords organized in an outline, spreadsheet, flowchart, or tree diagram. While the name "checklist" implies checking items off a list, that is only partially true with an auditor's checklist. The checklist is also a great place where specific objective evidence is being entered to show each requirement being met or not being met. This will make not only the audit more thorough, but also make writing the audit report so much easier.

Checklist Rules

1. Prepare before the performance phase
2. Link the question, statement, or keyword to the source of the requirement
3. Leave space for comments and observations

A technical approach to writing checklist questions is that they be *yes/no* and single-issue.

For example, if we created yes/no checklist questions for the mandatory requirements (shall, must, will) they may be:

- Does the organization conduct internal audits at planned intervals? (ASQ/ANSI/ISO 9001, clause 9.2.1, Internal Audit).
- Are obsolete and outdated packing materials (labels, labeling) destroyed? (GMPs, 211.122 (e), Materials examination usage criteria).
- Are there documented procedures for the validation of sterilization processes? (ISO 13485, clause 7.5.7, Particular requirements for sterile medical devices).

However, even though *yes/no* checklist questions provide excellent traceability to requirements, they can be ineffective if used as interview questions. If asked a yes/no question, the person you are interviewing may simply answer yes or no. Your interview questions should be open-ended (interview techniques will be discussed during the performance phase). In other words, the yes/no question on the checklist is for you to figure out either "yes" or "no" while/after you are interviewing and have obtained evidence to support your "yes" or "no" finding.

When creating checklists, keep in mind the items depicted in Figure 5.2. The checklist should be properly identified (page, version, title), include your referenced question or statement, and allow space for collection plans and recording observations. It is good practice to include checklist questions from at least two document levels (for example, ISO 14001 and department procedures).

When you go to the area to be audited, you will know exactly what to look for and listen to. As you are observing and listening to the people in the area explaining how they do their jobs, you are getting your checklist questions answered. This technique is thorough, traceable, and the key to successful and effective audits.

Standard: **Audit #:**

Ref.	Question or statement	Yes/ No	Comments/ notes
1.	Pull out the documents		***Document:***
2.	Select the control		Collection plan Items to examine
3.	Write questions/statements		Sampling plan
4.	Reference the requirement		***Record:***
5.	Repeat using at least two levels		Observations Interviewees

Version or last save date **Page 1 of X**

Figure 5.2 Checklist example.

If the auditee asks to see the checklist, it is normally okay to share blank checklists. However, it is not advisable to share checklists that contain your handwritten notes or your data collection and sampling plans. Knowledge of your collection and sampling plans would provide advance information concerning the audit evidence you plan to collect. Your completed checklist may or may not go in the final report, but there may be a requirement that it be filed with other audit working papers. So while some audit programs provide checklists as a matter of course, others keep them under strict distribution control for liability or regulatory reasons.

THE COMPLETED CHECKLIST:

- Provides structure and order
- Assures required coverage
- Provides opportunities for communication
- Is a place to record data/evidence
- Is a time management aid

Canned or pre-populated checklists distributed by your audit department or a consultant may not provide the flexibility that you need for a specific audit. Canned checklists work well for comparison purposes such as comparing different suppliers or operating organizations. When internal auditors are given canned checklists to use, they should study and understand the canned checklists before the audit. Internal auditors should also be ready to augment the canned checklist questions based on the controls being examined during the audit. See Appendix E: Example Checklist Page.

COLLECTION AND SAMPLING PLANS

You should determine what it is that you need to see during the audit (data collection) in order to verify that controls are in place and being followed. You should write it down and either put that information in the checklist or keep it separate. A collection plan is a list of the things you want to see, such as purchase orders, defective items, loading, an autoclave process, or inspection records. A sampling plan specifies how many and what samples you need to look at, such as the kind and number of purchase orders, defect item reports, truck loadings, autoclave batches, or inspection records. There is an old auditor joke that your sample sizes are determined by how much time you need to get to the airport. However, as an internal auditor with unique knowledge of your organization, you need to see enough evidence to satisfy that the requirement is being met. Remember that your industry or organization may require a specific sample size because of industry risk or perhaps the previous results of the area due to risk-based auditing practices.

Audit Principle

Ensure that sufficient samples (records, product, processes, interviews, and so on) are taken to match the purpose and scope of the audit.

Auditors must choose the samples they require unless it is a 100 percent examination. For example, if you need to verify that customer complaints are recorded and there were only three complaints this quarter, you can examine all three of them. On the other hand, if there were 100 complaints per month, you would only have time to look at a sample, such as a 10 percent sample. The rationale for the size of the sample you take should be addressed in your procedures. Auditors should randomly select the samples to be used as evidence. For a long list of items (for example, 200 items), ask the auditee for a number from 1 to 10 (for example, the auditee selects 3) and then check the third item for every group of 10 (for example, #3 of group 1–10; #13 of group 11–20, #23 of group 21–30, and so on) for a 10 percent sample of the total number of items. To get a quick random number, pick a thick book, thumb through the pages toward the back and then pick a page (for example, page 1258) and use the last (8) or last two digits (58) as random numbers as a basis for your selection.

Audit Principle

Samples must be random and representative unless specified objectives require otherwise.

For internal audits, there are several reasons for nonrandom sampling. Nonrandom sampling is called *directed* or *judgmental sampling.* An auditor may be directed to sample certain activities because of their importance or potential problems. An auditor may use their judgment to test historically weak areas. In internal auditing, sampling should be done in the most effective and efficient manner of benefit to the organization.

WORKING PAPERS OR WORKING DOCUMENTS

Working papers include checklists (discussed earlier), guidelines, log sheets, forms, sampling plans, flowcharts, and anything that will aid you in conducting the audit.

Working papers may not be papers at all. You can use electronic media, photos, or recording devices, as well as paper media. You may create your own working papers or they may be given to you by the audit program manager.

The are two basic rules for working papers:

1. Working papers must be flexible and not detract from the effectiveness of the audit. If the use of a form restricts an auditor from doing the best job possible, the form should be redesigned or deleted.
2. Working papers must be safeguarded. In some cases, such as sampling plans, working papers must be safeguarded from the auditee. In other cases where working papers contain sensitive information about an auditee organization, they must be safeguarded from outsiders.

Next, let's look at evaluating documents and determining the audit strategy for the upcoming audit.

Chapter 6

Desk Audit and Audit Strategies

Before the on-site portion of the audit, you must become familiar with the controls used in the area to be audited. The familiarization process could comprise (1) a formal document evaluation (desk audit) and report, or (2) reviewing documents or other documented information to add questions to your checklist, and/or (3) flowcharting processes to help in your understanding of them. Auditors should use various techniques to understand the system and processes they will be auditing.

DESK AUDIT/DOCUMENT EVALUATION

Auditors evaluate documents to ensure that the auditee's management system (controls) is adequate to meet higher-level standards or guidelines (Figure 6.1). You can sit at a desk or table and compare the auditee's documented management system to the requirements of the standard(s). Desk audits (documented information evaluations) normally are conducted when there is either a new requirement

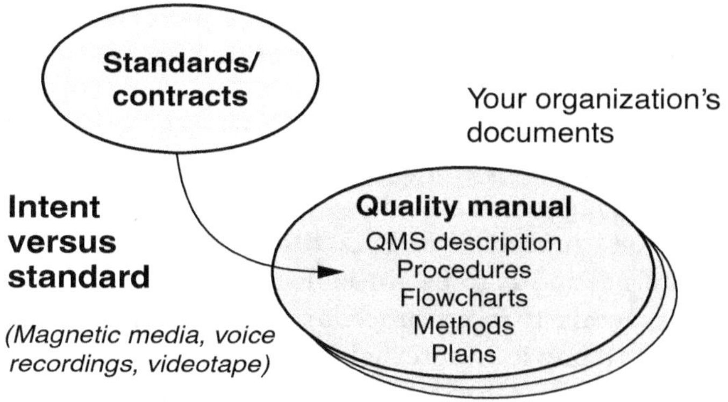

Figure 6.1 Determine the adequacy of the system.

standard or there are changes to the organization's management system controls. This is not a "remote audit," but it can be considered part of the overall audit to reduce on-site time (which can reduce the auditee staff being away from their positions). An actual remote audit may still take place later on, but a desk audit is basically taking place at the comfort of your desk or designated site such as at a Sensitive Compartmented Information Facility (SCIF). Incidentally, as an internal auditor (you are all on the same team) you can sometimes reduce on-site time by requesting certain documents and records to be sent back and forth through an inter-office system if your organization allows something like that. However, do not overly rely on what you can obtain in this review to replace the full experience of the audit.

To conduct a desk audit, you must first create a checklist (or acquire one from the audit program manager) of the higher-level standard(s) (that is, performance standards and contracts). Then for each higher-level standard requirement (such as ISO 9001 or ISO 14001) you check off where you found the requirement addressed in the auditee's management system.

Some requirements are very clear, such as a requirement for a procedure or ISO 9001 requiring the results of internal audits to be available as evidence. If the desk audit reveals that no procedure has been issued or no results of previous internal audits are available as evidence, there is a basis for nonconformity. Similarly, if there is a requirement for a record or review and there is no provision in the management system for a record or review, there is a basis for non-conformity.

When it is not clear that high-level requirements are addressed by lower-level documents (procedures, work instructions, plans, and so on), you must determine if the intent of the requirement is addressed. Then later during the performance phase, that intent can be tested and confirmed. Not all requirements can be verified with a desk audit because not all requirements require traceability to a controlled document (any medium).

There is a potential nonconformity if the requirement (intent) is not addressed in the organization's documented information. If you find several glaring nonconformities, there may be a reason to cancel or delay the audit. As an auditor, you cannot audit a system that does not yet exist. There must be a system or process, it must be implemented, and there must be records that the system has been maintained for a period of time. People characterize a new procedure by saying 'the ink hasn't had time to dry yet.' An auditor cannot determine if a procedure has been effectively implemented and is working unless it has been used for a period of time. Your organization must determine what that period of time should be. For us, 30 days is not enough; perhaps three months is sufficient. The bottom line is that if a determination is made that a specified control has been effectively implemented and it was just published, you can conclude only that the procedure exists. Such a situation could result in rescheduling the audit so as to not waste resources.

Audit Principle

Verify that there is an established system/ process to audit before the audit.

Using the checklist to evaluate documents and records prior to the audit may generate other questions to be answered during the performance phase. The desk audit report should identify any missing documents (required procedures, plans, and so on) or records.

The desk audit report can be a list of non-conformities referencing the requirement, or the checklist itself can be used to indicate (yes/no) requirements not addressed in the documentation. Other documented information that can be used to better understand the organization to be audited are prior audit reports, history of performance, and records. Resolve any concerns about the adequacy of the management system and/or process before you proceed.

FLOWCHARTING

A wonderful technique to help you understand the processes you will be auditing is flowcharting. You can use it to bring confusing procedures to light or to understand the key elements of the process you are about to audit. The purpose of a flowchart is to describe a process or system (for example, how work is performed). Flowcharting is like drawing a picture of the process. See Figure 6.2.

There are many different flowchart styles and techniques. Flowcharts can be constructed using pictures and/or symbols and put in horizontal or vertical arrangements. Flowchart symbols can be found in ANSI X3.5.

Flowcharting Tips

- You should limit the number of blocks so that your chart is not too complicated. If there are a lot of blocks, move to higher-level controls. For example, instead of showing the 12 detailed steps for a product transformation or service delivery, you simply block it as stamping, forming, finishing, and so on.

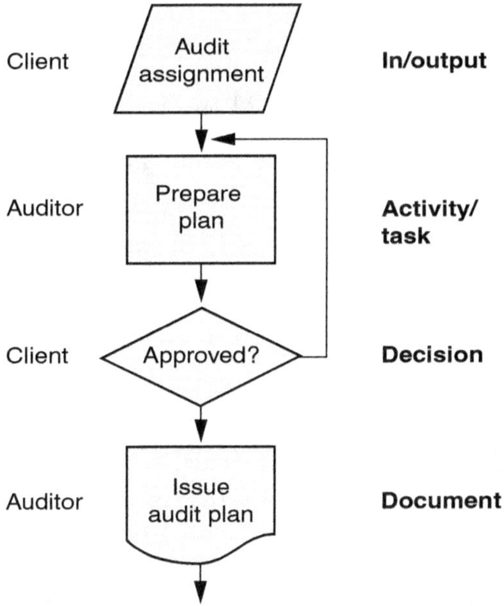

Figure 6.2 Basic flowchart symbols.

- For auditing, you will be most interested in checkpoints where a decision is made or should be made. Outputs of processes should meet predetermined criteria for you to know that the output is acceptable. If the output does not meet the criteria, something has to be done about it.
- Your flowchart should also contain who does what. This will help you know who to talk to about an activity or step.

- Add any documents and records required that relate to an activity.

A summary of flowcharting benefits includes that it:

- Provides information about the process steps and their sequence
- Helps with identification of problems/ improvements
- Is a valuable tool for training programs
- Makes it easy to identify checkpoints
- Makes it easy to identify responsibilities

Auditors should use various techniques to understand the system and processes they will be auditing. Besides the desk work (reviewing procedures and flowcharting), auditors can tour an area beforehand to better understand how things work.

AUDITING STRATEGIES

As an internal auditor, you may be assigned a process, area, function, or department to audit. Or you may be assigned a common element such as document control (or control of documented information requirement) or your CAPA process (or the Nonconformity and Corrective Action clause) to audit in an area or across several departments.

These are called *element* and *department* strategies:

- The *element* method is horizontal and is auditing according to the standard element. It is good for linkage to standards. The element method has been abused in the past and resulted in very narrowly defined audit scopes that do not include inputs, outputs, and other process linkages. New auditing approaches suggest that the element method should be limited to common system elements such as the corrective action process or document control.
- The *department* method is vertical and is auditing according to each department or function. It is good for accountability and following the process flow.

When you audit by department or element, you can use tracing techniques to examine the controls. See Figure 6.3. *Tracing* is used to follow the path of a process (procedure or method) to test out controls. You can trace the process forward or backward. As you trace, you can ask questions about the procedure or process and get your checklist questions answered.

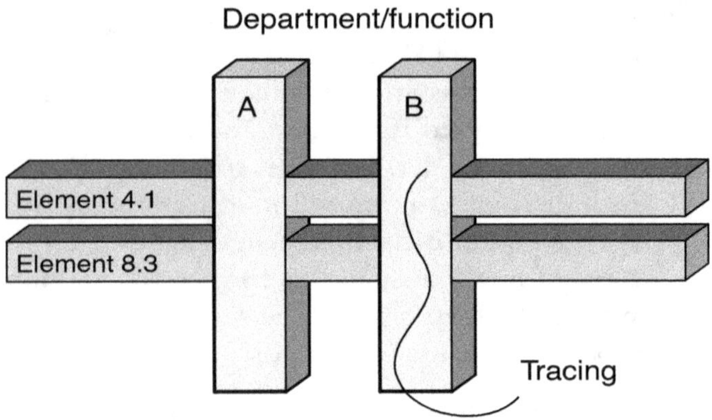

Figure 6.3 Auditing strategies.

For example, tracing will work when information is needed on how a document gets changed, how a part is finished, or how a service is conducted.

If you are uncertain of conformance or nonconformance, tracing can be used to verify that the requirements are either addressed or not addressed by the auditee. Tracing may take you to other departments to verify an input or output of the area you are auditing. However, be sensitive to the perception that you may be moving outside the agreed purpose and scope. If necessary, assure the auditee that good audit practice necessitates that inputs, outputs, and process linkages are tested and verified.

Audit Principle

Stay within the agreed scope unless risk necessitates other actions.

Once the scope is set, you should stay within the scope and use your judgment when problems are found outside the scope. An auditor has an obligation not to ignore problems found outside the original scope. This does happen from time to time. Varied problems can be observed at the audit as you are working. From uncovering a major safety issue not readily seen by others to casually noticing a minor issue that the auditee should address in general (e.g., liquid spill or missing tape)—you are all on the same team. In other words, problems need to be addressed even if not within the scope.

Method for handling problems outside the agreed scope of the audit:

1. Determine if the problem is potentially harmful (e.g., safety concern uncovered or damaged product noticed in shipping area prior to shipping, or possibly an open door that should be secured) or minor (e.g., nonconformity outside of scope or maybe a spill in the hallway, or parking lot filled with pallets with no place for parked vehicles).

2. If minor, report to the auditee and continue auditing within the scope.

3. If potentially harmful or a major issue, report to the lead auditor (audit program manager if you are the lead auditor) and auditee management. Determine whether the situation warrants further investigation and whether the audit should be stopped or continue within the original scope.

4. Report major problems found, although it is not necessary to put them in the audit report as non-conformities. The audit report will usually have room for a general summary in which serious issues uncovered outside the scope can be included and would assist in the tone of the report.

Finding problems outside the scope that require your immediate attention is unusual, but you must be prepared for it. Certainly, events that could result in personal injury, hazardous spills, fire, or shipping of defective products require immediate attention. Auditee management may scrutinize how you handle yourself in such situations.

This concludes your preparation. The next chapter cover the performance phase of the audit, which is your opportunity to dialogue with people and collect objective evidence.

Chapter 7

Beginning the Audit

In the last chapter you made your final preparations for the audit. Now you are ready to start the performance phase. You are ready to collect evidence to verify that people are complying with external standards and internal procedures. The opening meeting represents the start of the performance phase and establishes the official communication links between the audit team and the auditee.

OPENING MEETING

The lead auditor or audit team leader is responsible for the opening meeting. As lead auditor you will need to determine how formal the opening meeting should be. Generally, opening meetings for internal audits are less formal than opening meetings for external audits.

The purpose of the opening meeting is to confirm the audit plan, team introductions/roles, verify communication channels, confirm the summary of planned audit activities, and answer auditee questions (ASQ/ANSI/ISO 19011, clause 6.4.3).

You should always schedule an opening meeting. Even if this is a routine system audit, it is common courtesy to let everyone know that the audit team is in the area and what your plans are for the audit. If the audit is routine and everyone knows what to expect, you can keep the meeting short. A short meeting may be held in the supervisor's office and take less than five minutes.

A more formal meeting should be held for larger audit scopes when the audits are not routine, or when the business risks are higher. A formal meeting may be held in a conference room and take 30 to 60 minutes. The meeting ensures that everyone is aware of the audit and allows any last-minute issues to surface. If it is an audit of a new area or there are new people involved (new to the audit process), then expect the meeting to take longer.

You should keep a record of who attended the opening meeting. Some auditors pass out a sign-up sheet (name, area, date). You should also record any audit plan changes or concerns of the auditee. The agenda items in the next section should be key discussion points. (See Appendix C: Example Meeting Agenda and Record.)

OPENING MEETING AGENDA

Complete Introductions. Make sure everyone knows each other. This is an ideal time to take attendance.

Thank Your Host. Thank the person (or acknowledge him or her) who made the arrangements for the audit. This can be anyone who coordinated the audit.

Review the Audit Plan. Reaffirm the purpose, scope, and standards to be audited against. If corrective actions from prior audits are to be verified as part of the audit, this should be in the purpose, too. You should clarify any unclear details of the audit plan.

Limited Access. Any accessibility limitations placed on the auditors should have been identified before the opening meeting, but be prepared to address any last-minute issues. The auditor's access to certain areas may be limited for several reasons (Figure 7.1).

Normally, accessibility is not an issue in internal audits. However, security and a need-to-know basis for access have become more important in today's business climate in more industries.

Safety restrictions are common. Always comply with all safety and environmental rules. As with the law, ignorance of safety or environmental rules are no excuse. Ensure that you have proper training and personal protection equipment and know how to use it. Always follow the posted guidelines even if others are not, or a manager says not to worry, or if the area you are walking through is not even being used for any reason. As the auditor, you are being watched and word travels fast that you are too good for the posted guidelines.

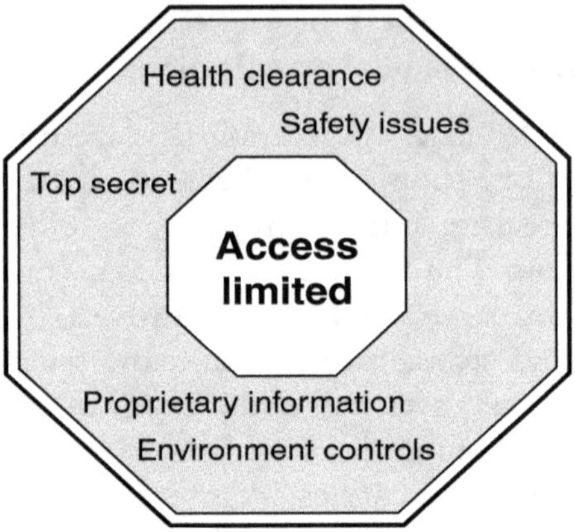

Figure 7.1 Reasons for limited access.

> **Audit Principle**
>
> *Comply with auditee rules (safety, environmental, health, restricted areas, and so on).*

Audit Methods and Techniques. Explain how data will be collected, such as through review of records and other documented information, observations, and individual interviews, and if any portion of the on-site audit will be remote or virtual.

For mature audit programs, it may not be necessary to cover this agenda item for every audit. You may simply ask if there are any questions about how the audit will be performed. Be prepared to explain your approach to sampling (that is, random or directed). If you are likely to audit more than one area using process or tracing techniques, explain that too.

Reporting Process. Explain how the data collected during the investigation will be reported and followed up. The audit findings may be reported as conformity and/or nonconformity (noncompliances, findings, remarks, observations, or improvement points). Explain how the relative importance of results is categorized, such as major and minor nonconformances or other grading methods used for your organization. We will discuss more about grading nonconformities later on, after the performance phase. As an internal auditor, there should be less worry about any sort of actual grading of nonconformities and more concern that a nonconformity was uncovered. That said, some nonconformities will obviously be more important than others and the internal auditor should order the nonconformities found in order of importance, regardless of any grading methods. For routine audits, everyone should already be familiar with the reporting process.

Establish the Interview Schedule. For routine audits where everyone is expected to be available for the auditor, the schedule may simply be a time (interviews from 9 am to 12 pm). However, most organizational cultures require formal interview schedules. Be sure to follow your organization's guidelines. Confirm the availability of personnel (interviewees), and resolve and record schedule changes or limitations. (See Appendix D: Example Interview Schedule.)

Review Logistics. Verify meeting room locations and a home base for the auditors with necessary equipment and services (electrical power outlets, restrooms, telephones).

Confirm the Exit Meeting. The exit meeting is very important and deserves special mention. Confirm the date and time of the exit meeting and who will be attending. You should also verify the times of any interim meetings.

How you handle yourself and your presentation techniques in the opening and subsequent meetings will have a significant effect on setting the tone for the audit (auditing attitudes). An audit, whether internal or external, is always serious. Internal audits may be less formal, but the process of interviewing, probing, and examining to judge conformance or nonconformance should be done in a cordial, businesslike manner. Some new auditors make the mistake of trying to be casual or even flippant with their comments to reduce auditee stress. However, this comes off as being unprofessional.

The audit team should meet with the department manager, supervisor, or area coordinator who arranged for the audit. Exactly who attends the opening meeting may depend on the organization's culture and upcoming events. If the organization is due for a visit from a regulator or certified/notified body, managers may use the audit experience to prepare their personnel.

If the auditee provides escorts for the auditors, the escorts should be at the opening meeting, too. Many internal audit programs don't require internal auditors to be escorted but there are exceptions. Company proprietary issues and organizational culture may indicate the need for escorts.

If an escort is provided, he or she may perform the duties listed in Figure 7.2.

Sometimes senior management attends the opening meeting to show support for the audit program or because they are deeply concerned with the performance of the area to be reviewed. Your organization may have guidelines for opening meetings that need to be followed.

- Make personnel introductions
- Clarify information when asked by the auditor
- Keep management informed of progress
- Be the auditor's guide
- Confirm or deny nonconformances
- Ensure that auditors comply with rules (safety, environmental, health)

Figure 7.2 Auditor escort duties.

In addition, the lead auditor can also:

- Share the checklist with the auditee (if not sent earlier)
- Identify needed documents or records to be supplied by the auditee
- Explain how improvement areas will be reported, if at all
- Identify any union-management issues

At the end of the meeting, the lead auditor should ask for any questions or items that need to be clarified. For routine internal audits, you may only need to let the auditee know you are ready to start, confirm the interviews, and establish a report time.

> **TIP**
>
> **Meeting time is not audit time. You are not collecting data to verify conformance while you are in the opening meeting. Keep meetings short, don't let the auditee take over meetings, stay focused, and get busy auditing.**

OTHER MEETINGS DURING THE AUDIT

If the audit lasts more than one day, you should schedule daily meetings to keep the auditee informed of your progress. You will also need to schedule audit team meetings to coordinate the audit. The timing of meetings is at the discretion of the lead auditor. Meetings should be as brief as possible.

> **Audit Principle**
>
> *Keep the auditee informed of audit progress.*

Agenda: Audit Team Meeting

- Share data/evidence/information
- Re-plan assignments
- Review and record observations
- Determine compliance
- Start the reporting process

Agenda: Meeting with the Auditee

- Verify areas completed
- Confirm areas still to be completed
- Identify problems uncovered

> **TIP**
>
> **If the auditee claims to be too busy for an audit progress report, find another means to keep the auditee informed. Other means include: voice or e-mail, hallway encounters, short notes in a mailbox, and so on.**

WORKING PAPERS

Auditors may use several different forms and documents (called working papers) to help them perform the audit. Working papers may be provided by audit program management or created by the auditor.

The following are examples of working papers that you may encounter:

- Audit procedures
- Memory joggers
- Forms

- Sampling plans
- Auditee evaluation forms
- Attendance record form
- Audit questions
- Log sheets
- Guidelines

The working papers represent a place to record data and to provide guidance during the audit. (See Appendix E: Example Checklist Page.)

The meetings are over and it is time to gather audit evidence. Next, we will discuss how to interview people and collect data.

Chapter 8

Data Collection

The purpose of the performance phase of the audit is to collect audit evidence. The audit evidence collected determines conformance or nonconformance.

Your job is to collect factual evidence of conformance to requirements. Requirements are found in standards, procedures, and other documents listed in the audit plan. The requirements you audit against are called the audit criteria.

The vast majority of audits are conducted to determine the degree of conformance to national or international standards and organizational documents (policies, procedures, instructions). You should collect data (evidence) according to your collection plan.

COLLECTION PLAN

You will need evidence from:

- *Documents and records and other documented information.* Review procedures and examine records.
- *Physical examination.* You count it, it is tangible.
- *Observing activities.* Watch what is going on.

- *Interviewing.* Talk to people connected with the process.

As part of the preparation for the audit, you reviewed documented information (procedures, flowcharts) that described the system to be audited. You should have made a note in your checklist or data collection plan of things that can be checked to verify an activity. During the audit, you may discover additional items that can be checked and they should be noted, too.

When reviewing documents and documented information, look for where promises are made. In particular, note promises that link with higher-level standard requirements. For example, promises to follow or issue a schedule, complete a record, file a form, assign certain personnel, create and maintain an environment, use specified equipment, report within a certain time frame, check off certain tasks, and so on.

EXAMINATION OF DOCUMENTS AND RECORDS

Documents

Prior to the audit, documents and other documented information were evaluated to determine the adequacy of the system and used to develop checklist questions. Using international standards lingo, documents are considered part of your organization's total documented information. During the performance phase, documents may again be referenced to verify process steps or the proper sequence of activities. Documents can be procedures, manuals, policies, or work instructions. Documents specify what should be done.

Documents should be checked to:

1. See if rules exist
2. Compare them with actual practice (important to verify what is intended versus what is actually being done)
3. Better understand the auditee's operation or business

Records

Records can be thought of as specifying what has been done. Some industries view records as factual history, so without evidence of the record then the activity "never happened" and is cause for a citation. Checking records is one way to verify that performance standards are being followed. Verification of requirements through records provides a very high level of confidence in compliance. People don't normally falsify records, and if they do, they are typically subject to severe penalties. Using international standards lingo, records are also considered part of your organization's documented information.

Audit Principle

When an unethical activity is observed, verify it, record it, and report it.

Verify that records are:

- Being completed
- Sufficient for evidence of conformity

Verify that document and record controls are:

- Current and available to users
- Approved, identified, legible, and maintained

A typical and effective way to verify controls is to flowchart a procedure and then to trace the actual steps of the procedure, all the while looking at records, interviewing people, observing work, and collecting physical evidence (an example of process auditing).

Documents and records can be in any medium, such as electronic or paper. If performance standards call for document and records control, there may be additional requirements for approval, removing obsolete documents, establishing retention times, and so on.

INTERVIEWING PEOPLE

Interviewing people may be the most challenging and rewarding part of the audit performance phase. Some auditors may view interviewing as a contest between the auditor and auditee with the auditor trying to find nonconformances and the auditee trying to hide them. That is the wrong approach and will promote conflict. You should remind yourself that you are on a fact-finding mission and the interview is just another opportunity to get the facts.

Many consider the interview as the most difficult part of the audit to do effectively. Dealing with people is always more of a challenge than dealing with inanimate objects. Interviews provide very valuable information that you may not be able to learn by other means. However, interviewee statements are not as reliable as a record. Interview information to be used in the audit report should be *corroborated.*

Corroboration or verification can come from:

- Another person
- Observation
- Documents and records
- Another auditor

For second- and third-party audits, information should always be corroborated. For internal audits, you can normally accept an admission of guilt (I forgot to complete the record, I bypassed the approval step, and so on) without seeking corroboration. If you have a question about your policy, check with the audit program manager.

Being an effective interviewer requires assertiveness skills. If you feel this is a personal area that can be improved upon, you should consider taking an assertiveness class. Both aggressive and nonassertive (passive) auditor behaviors will result in ineffective interviews.

While interviewing, note when the interviewee uses the words *normally, most of the time,* or *usually.* These are red flags for you to ask about what happens when it is 'not normal.' The best processes function well even when things are not normal or during a crisis.

One-on-one, face-to-face interviews are preferred and usually the most effective (see Figure 8.1). When interviewing more than one person at a time, one interviewee may start answering for the other or the interviewees may team up against the auditor. If the auditor does not take back control of the interview, the interview information may be worthless. If the auditor takes back control of the interview in an abrasive or aggressive manner, the interviewees may become defensive or hostile. However, there may be times when group interviews are appropriate. For example, you may want to interview an entire team to encourage team building and reduce individual stress.

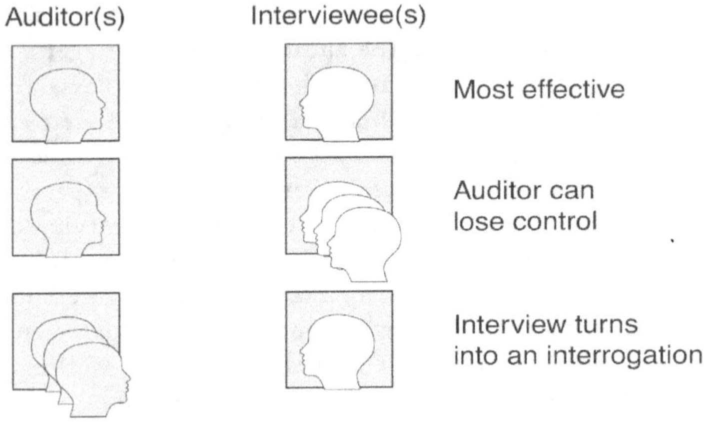

Figure 8.1 Comparison of interviewing methods.

When multiple auditors are interviewing one auditee, the auditee can become defensive or overwhelmed. If you have a second person with you on the audit interview, you should explain why he or she is there. It may be that the second person is there to take notes, be an observer, serve as a subject matter expert (technical specialist), train, or act as another auditor who will be asking questions regarding different criteria. When multiple auditors are interviewing one person, they should be very courteous and aware of over-pressuring the interviewee. Some auditors can sense changes in interviewee moods and attitudes and adjust for them.

A *six-step method* for interviewing, popularized by Dennis Arter, is a commonly accepted practice. Before starting the interview you should remind yourself that you are a guest in someone else's area. At first, try to put the interviewee at ease. You may need to discuss the weather or a national news item to lower the interviewee's anxiety. Be polite, shake hands, introduce yourself, and explain why you are there.

Six-step interview method:

1. Put the person at ease.
2. Explain your purpose.
3. Ask what they do.
4. Analyze what they said.
5. State your conclusions.
6. Explain your next step.

It is during step 3 that you can get your checklist questions answered. Be sure to take good notes and keep a record of the responses.

Interview guidelines:

1. Interview questions should be open-ended (for example, ask, "What is the role of your function?" "What do you do?" and so on).
2. Ask to see the records, documents, or other means to verify controls.
3. Listen; don't talk except to ask questions or paraphrase answers.

TIP **Never lecture the auditee. When you are lecturing, you are not collecting data. Secondly, the auditee is not interested in your views, he or she only wants to know if the area passed or failed.**

It is not considered good practice for an auditor to ask yes/no questions in an interview unless you are specifically using that type of question as a technique to calm a person or to refocus on the topic. There are times during an audit when the auditor needs a yes/no verification, such as "Are you maintaining the records or not, yes or no." However, getting yes/no answers will not give you any additional data about how the requirement is implemented, the person's knowledge about the requirement, or where to go to gather additional evidence.

Communication problems (between the auditor and interviewee) are probably the principal difficulty that must be overcome during an audit. If you think you might benefit from some communication pointers, consider taking a course on improving communication skills.

PHYSICAL EXAMINATION

Physical examination concerns the tangible, things you can count or measure in some way. It is the most reliable source of objective evidence. Numbers are generated. If you use measuring equipment, the equipment should be accurate and be under calibration control.

Examples of physical evidence:

- There were 12 items in the nonconforming bin.
- The three trucks in the yard passed the weight test.
- The check scan confirmed the original scan.
- All packages on the dock complied with regulations.

OBSERVATION OF ACTIVITIES

Observing is using your senses (Figure 8.2). You may look around, be aware of smells that may be improper (chemical release), listen to people and the work area sounds, and in some cases even touch and feel something (for example, is the spot wet or greasy? Is it rough or smooth?).

You can observe processes to confirm implementation and ongoing maintenance of the system. It is best to observe an actual task being performed rather than a practice run or one that was created for you (the auditor). At the same time, avoid interfering with the performance of the activities. If you do interrupt or redirect the process, be aware of the artificial influence being created. If you sense or observe that an operator is nervous, take time to put him or her at ease and bring about a return to a normal work environment before you proceed.

Figure 8.2 Use your senses to make observations.

VERIFICATION AND VALIDATION

Auditors collect evidence to ensure that requirements are being met. Auditors may verify and/or validate that requirements (audit criteria) are being met. In general, verification is checking or testing and validation is actual performance of intended use. The dictionary does not support the distinction normally drawn between verification and validation in management systems and system/process audit standards. We need to draw on the verification and validation definitions provided in ASQ/ANSI/ISO 9000 and the design and development model outlined in ASQ/ANSI/ISO 9001, clause 8.3.

Verification

Verification should be performed to ensure that the system/process outputs meet the system/process requirements (audit criteria). Verification is the authentication of truth or accuracy by such means as facts, statements, citations, and measurements, all of which are confirmation by evidence.

An ingredient or element of verification is that it is independent or separated from the normal operation of a process. The fact that an auditor is checking whether the process or product conforms to requirements is itself verification (as opposed to inspection checks).

For example, ASQ/ANSI/ISO 9001, clause 8.3 Design and development, requires consideration of verification by comparing designs to similar (but independent) proven designs or performing alternate (independent) calculations to verify the same results.

The most common method of verification is examination of documents and records. Records verify that a process or activity is being performed and results recorded. Interviewing is another method to verify that processes meet requirements via affirmation by the interviewee.

Validation

Validation should be performed to ensure that the system/process outputs are capable of meeting the requirements for the specified application or intended use. Validation is the demonstration of the ability of the system/processes under investigation to achieve planned results. Sometimes an activity cannot be verified by records or interviews and the actual process must be observed as intended to be operated. The observation can be the real process or a simulated one (depending on cost and practicality).

Some activities can only be verified because it may be too costly or impractical to validate a process such as a plant shutdown or start-up or the use of emergency procedures. Sometimes products or activities are only verified because the product would be destroyed or the process ruined by validating it (such as checking the seal on a container).

In another example, the auditee explains that a computer program automatically determines product markings and notices. The auditor may ask the auditee to submit a couple of products to view the selected markings and notices and compare them to requirements.

Many processes are required to be validated, such as sterilization. Auditors must ensure that the validations and revalidations are being carried out properly.

CONCLUSION

As you go through your checklist, match up audit evidence with every requirement. The existence of audit evidence is proof that the area under review has:

1. *Adequate controls* to meet requirements
2. *Implemented* and maintained the controls

Stay alert during the entire audit. By the end of the audit, you will be mentally drained from trying to assimilate all the data and how it relates to the audit criteria (requirements).

Sometimes collecting evidence to verify conformance is not very straightforward. In those situations, you will need to apply other techniques, such as the process technique in the next chapter.

Chapter 9

Techniques to Improve Effectiveness and Address Vague Requirements

Auditors need to be able to employ several auditing techniques and strategies to accomplish the audit objectives as well as improve the effectiveness of the audit. It is difficult to verify conformity when requirements are vague or open-ended. Auditing by element or clause provides good traceability to requirements but can leave the linkages between processes untested. Auditors may encounter situations where there is no documented procedure yet must determine if the process is controlled and conforms to requirements.

In the absence of prescriptive requirements, auditees must still demonstrate to the auditor that they conform to requirements. This chapter is about approaches for verification of conformance to open-ended requirements and using process techniques to test the management system linkages. For the auditor, all requirements must be verifiable and traceable.

CLOSED-ENDED REQUIREMENTS

Most standard clauses contain very specific prescriptive requirements. We can think of prescriptive requirements as being closed-ended because they are very explicit. For example, if a standard requires a procedure, the auditee must have a procedure. If a procedure requires a red stamp, the auditor expects to see a red stamp.

For auditors, closed-ended requirements can be listed and checked off with a yes or no answer (on the checklist). The user creates the record, procedure, or plan and the auditor checks off his/her corresponding observations. Closed-ended requirements are easy to check and are traceable. However, standard writers don't write or style a standard for ease of auditing. Standards and procedures are likely to contain vague and nonprescriptive requirements that may be difficult to confirm. Auditors must not avoid or gloss over verification of any requirement but must employ techniques to verify conformance to all requirements. If you ever encounter a requirement that you cannot verify and is not auditable, report it to the audit program manager and client.

OPEN-ENDED REQUIREMENTS

Some standard clauses and internal organization procedures may have open-ended type requirements that are not very specific and can leave the auditor with a lot of questions. You may notice various open-ended requirements during the document evaluation and during the performance of the audit. Open-ended requirements are very popular for internal procedures and instructions (to provide maximum flexibility). You may have heard someone say, "That requirement has holes so big, you can drive a truck through it." That may be the case, but the auditor still wants to know what kind of truck, how fast it is going, whether the driver has a license, and so on.

I have identified four types of open-ended requirements you may encounter during your audit. These are shown in Table 9.1.

Table 9.1 Types of open-ended requirements.*

Type I: Open-ended phrases/words
Use of open-ended words subject to wide interpretation. Words such as "periodic," "timely," "readily," "promptly," "without undue delay," and "based on importance" are not definitive. "Periodic" indicates repeatability but no frequency. "Timely" is relative to other undefined factors occurring concurrently or in the recent past or future. "Importance" is relative to the units being compared against.
Type II: Generalized statements
Phrasing a requirement at a generalized or abstract level (for example, to control a function or process).
For example: The organization shall control planned changes. The organization shall implement production and service provision under *controlled* conditions.
Type III: Unclear or undefined words
Use of words that are not defined or are subject to multiple definitions, which can leave the auditor with no basis for issuing a non-conformance. For example: The output of this planning shall be *suitable* for the organization's operations. The organization shall determine the opportunities that need to be realized to *enhance* desirable effects for improvement. The organization shall *exercise care* with customer property.

* Accessed from ISO 9001 Transition Web-based training by the JP Russell Learning Center (www.JPRLearning.com).

Table 9.1, continued

Type IV: Goal but no tangibles specified
A requirement lacking specified verifiable actions or outputs (that is, there is no requirement to define, document, record, schedule, review, and so on). When there are no prescriptive requirements to audit against, audit findings could be perceived as subjective.
For example: Top management shall demonstrate leadership and commitment with respect to the quality management system by promoting improvement.
There is no requirement for a procedure or record or for top management to control the process.

Type I and III Discussion

Open-ended phrases or vague (Type I) requirements normally are clarified by a registrar, regulator, or the organization being audited. For example, *periodic* management reviews may be annual, or *timely* corrective action may be within 30 days. The planning of audits based on the importance of the process may be taken to mean auditing all ISO 9001 clauses annually. When interpretations are agreed upon (between the auditing and auditee organizations), auditors are bound to audit against the interpretations.

Use of unclear or undefined words (Type III) in requirements causes problems from time to time. This is a historical problem that is being addressed by guidance, standards, and interpretations committees or including definitions in standards such as ISO 14001, ISO 19011, or IATF 16949.

Additionally, standards such as ISO 9000 have been written to address word definition issues for the entire ISO 9000 family of standards. The main problems are that there are still some words that need to be defined and auditors don't study the available word definitions to better understand the requirements of the external or internal standard.

To audit the type of open-ended requirements found in Types I and III, auditors should seek additional guidance. The guidance could come from researching other standards (for example, the ISO 9000 vocabulary standard) and guidelines or from the auditing organization documents. Also, the application of some words may vary from industry to industry or area to area. A requirement to be prompt may be applied differently in the medical field or nuclear industry than for a soap manufacturer or boat company. If word definitions are a problem, auditors should seek guidance from their audit organization management.

In the absence of other guidance or regulatory requirements, an auditor should ask the auditee for their interpretation and audit the organization against it. For example: What is *timely*? What is *without undue delay*? What is *an acceptable planned interval*? Organizations may set periods or agree to a time on a case-by-case basis. You can then audit them to see if they are doing what they said they would do.

Type II and IV Discussion

Type II requirements for the managing and controlling of processes are very general. These types of requirement statements make perfect sense. It is only when an auditor must prove the negative (issue a nonconformity) that guidance issues surface. When is there lack of control? When is a process not being adequately managed? What evidence will withstand the scrutiny of the exit meeting and a subsequent review if a nonconformity is contested? Auditors want to be right the first time and not withdraw a nonconformity or noncompliance once they have determined one is justified. It is in everyone's best interest that the basis for nonconformity is clear and does not appear to be a subjective opinion.

Setting a goal (Type IV) as a requirement is useful if properly audited. Type IV open-ended requirements do not require the organization to manage or control and have no specific auditable requirements. Verification of conformance to Type IV requirements is challenging for auditors and audit organizations. This is particularly true for traditional compliance assessments where supplemental guidance may be appropriate. When Type IV requirements appear, auditors must challenge the auditee to explain how they comply. How do they demonstrate that they are promoting improvement? Or how do they ensure that equipment is maintained?

PDCA TECHNIQUE

To audit open-ended clauses, you can verify that the organization conforms to the intent of the requirements of the standard by using plan–do–check–act (PDCA) techniques (Figure 9.1). The auditor must seek to determine the existence of a process, how it was planned and implemented, and its outcomes. You can use PDCA to examine how the auditee addresses open-ended requirements and verify the existence of control when no documented procedure exists.

> Adequate control exists when an organization does the following:
>
> ***Plan.*** A plan, procedure, or method is developed (establish what needs to be done)
>
> ***Do.*** The plan, procedure, or method is being followed (do what was planned)
>
> ***Check.*** The plan, procedure, or method is monitored and/or measured against criteria
>
> ***Act.*** Action is taken to resolve the differences between expected and planned results (analyze and adjust the process)

Figure 9.1 PDCA technique for auditing.

You should seek answers to the following questions for the less prescriptive Type II and IV clauses in assessing conformance to requirements:

- *Is there a plan or method for conforming to the requirements? What is it? Has it been established?* Evidence may include an outline, plan, method, flowchart, markings in a work area, procedure, work instructions, or specifications. *Has it been implemented?* Evidence may be the existence of records, corroboration by interviews, observations, and so on.
- *Are there planned results (criteria)? Have they been achieved?* Evidence may consist of trend diagrams, recorded results, bar charts, matrices, comparisons, and so on.
- *Does the organization/person act on the results (make adjustments)?* When the output does not match the acceptable criteria, action should be taken to remedy the situation.

Common process interview questions:

How do you know what to do?	Verifies the existence of a predetermined method/plan
Tell/show me how you do it.	Verifies training and competency/knowledge
How do you know it is done right?	Verifies that acceptance criteria have been established
When it is not right, what do you do?	Verifies that action is taken on the results

The PDCA technique is a very powerful method to test all processes. You can use this technique in every interview where someone is assigned a job or task.

PROCESS AUDITING TECHNIQUES

There are process auditing techniques, process audits, and the process approach for management systems. This has been somewhat confusing to some, but the introduction of the process concept into management systems and auditing has been beneficial. In this book we do not discuss the process approach for designing, implementing, and maintaining a quality, environmental, or safety management system. In this section, we will discuss process auditing techniques that may be used in a process or system audit. The PDCA technique discussed in the prior section would be considered a process auditing technique. Tracing steps as outlined in a procedure or flowchart would be considered a process auditing technique as well.

The definition for a process is a series of steps that leads to a desired result or transform inputs into outputs. Process auditing techniques involve simply auditing the steps or activities and testing the linkages or handoffs between processes.

A process audit using process auditing techniques is an evaluation of the sequential steps and interactions of a process within a system. For example, an auditor may use process audit techniques during a management system audit of the purchasing or quality control department.

By its very nature, process auditing implies an action such as transforming inputs into outputs. Process auditing consists of evaluating the steps and activities that create the action or transform the inputs into outputs. This is a very useful technique because it focuses on the work cycle and deliverables instead of isolated requirements or controls.

The process model depicted in Figure 9.2 shows inputs, outputs, sequential steps, and a feedback loop for control purposes.

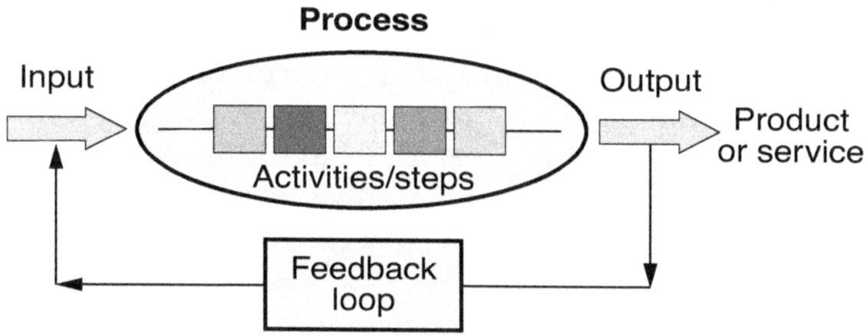

Figure 9.2 The process model.

Auditing a process or system using process techniques verifies conformance to the required sequential steps from input to output. Auditors use models and tools such as simple flowcharts, process maps, or process flow diagrams. Flowcharts typically identify inputs, people, activities or steps, measures, and outputs. The auditor normally gets this information from a procedure or flowcharts provided by the auditee organization. In system and complex process audits where the process cannot be followed in real time, requirements may still need to be verified. A technique to use to link processes within a system is for the auditor to record current customer names, customer order numbers, purchase order numbers, routing numbers, and project numbers during the first part of the audit (perhaps during a tour of the area to be audited) so he or she can link and verify process steps during the audit. For example, it would be impossible for any auditor to follow the requisition request from supplier selection and approval to issuing the purchase order, accepting the delivery, receiving inspection, and use in operations. However, during a tour of operations you could get the purchase order number of material being used, ask purchasing to see the requisition request, supplier approval, and purchase order records, then check inspection records in QC, verify on-time delivery, verify that supplier performance is being monitored, and so on. The use of process auditing techniques is more powerful and effective than auditing purchasing this month, QC next month, and shipping next quarter. By using process auditing techniques, you can test the linkages and communication issues between functions and groups.

The use of process techniques is a natural stepping stone from conformance to performance auditing. When collecting evidence, auditors will also observe performance issues that would be of value to management.

Auditors should report process performance issues that indicate potential for improvement. These indicators include:

- Waiting
- Redoing
- Deviating
- Rejecting

- Traveling excessive distances

All process performance findings should contribute to improvement programs such as lean, ISO 9001, or Six Sigma.

Most organizations are still auditing a process or a group of processes by element or clause and missing out on the value of process auditing techniques. The use of process auditing techniques provides added value by evaluating how processes flow, their controls and risks, and the achievement of objectives.

Auditors and management can benefit by using process techniques to better test and evaluate system controls. For more information on process auditing, check the ASQ website (www.asq.org) for process auditing books, web-based training classes, and instructor-led classes.

PROCESS APPROACH FOR MANAGEMENT SYSTEMS

The "process approach" or "system approach" is a business strategy for top management. They need to manage the product and service processes that add value to the organization. Most organizations started by being focused on their value-added processes. The salesperson (usually the owner) got the customer's order, he gave it to the shop guy, they talked and planned, and the shop guy made it and turned it over to the dock person to ship to the customer. As organizations grew, matured, and became more complicated, functions specialized. Top management lost sight of the purpose of the organization. Management focus turned to engineering, marketing, production, accounting, and so on. Communication barriers were created and some organizations became dysfunctional. The engineering department was one of the best in the country but when designs were handed over to production, they were a disaster, Marketing could not be beaten but sold what they didn't have, and Production set new records for a model that did not sell, and so on. Departments became experts at ensuring they were not blamed for doing the wrong thing instead of working together to do the right thing.

Quality tools such as process maps and the like are of no value unless organizations refocus on value-added activities. Top management should bring together all functions involved in a process to focus on common objectives. Meetings should not just be about what was done in the past, but how to work together so the organization wins. An already well-managed organization does not need to stop doing what works because of the process approach; it just needs to learn from the new tools and techniques.

After collecting your evidence, you will need to figure out what it means. Next, we will discuss how to analyze the data before the exit meeting.

Chapter 10

Analyzing the Results

Now that you have completed the investigation and collected evidence, it is time to analyze the information. You may analyze the audit evidence on your own, with your team, or both. Recall the four types of evidence that we discussed earlier (the documented information of documents and records, interviews, physical, and observations).

There should be evidence to verify conformity or nonconformity to requirements. There should be sufficient audit evidence to fulfill the purpose and scope of the audit. If there is not sufficient evidence, you should continue to audit or report any limitations or contingencies at the exit meeting or to the lead auditor.

Any contradictory evidence or unresolved issues should be resolved prior to reporting the results of the audit. If not, either the report should be delayed or the unresolved issues should be made clear to all parties (client, auditee, auditors, and so on) at the exit meeting. However, it is highly undesirable to leave issues unresolved or inconclusive unless there are special circumstances (cause of the unresolved issue). You can request additional information that may result in a supplemental report later on after the new information is analyzed.

CLASSIFICATION OF OBSERVATIONS

The evidence that you collected before and during the audit must be examined (analyzed). The data may be recorded on a checklist, in a log (a record of auditor's observations), or seen in a photograph, on notes on blank forms, or in references to auditee documents and records.

A datum is considered objective evidence if it can be proven true and is free of bias. It can be proven true if it is traceable (to verify) or reproducible (another auditor could collect the same datum).

> **Audit Principle**
>
> *Conformance and nonconformance must be verifiable and traceable.*

How you sort the data should be consistent with your organization's guidelines for classifications used by your organization to report results. Most of the time, results are in the form of a non-conformity statement, as a violation of a specified requirement. Reporting audit results as nonconformity statements (as opposed to other techniques) is a very effective tool for implementing and maintaining a management system and monitoring conformance to a particular standard or contract.

The next step is to sort the data based on importance (significance) and relevance (Figure 10.1). Is the information relevant to the organization being audited? Has a requirement been broken (violated)?

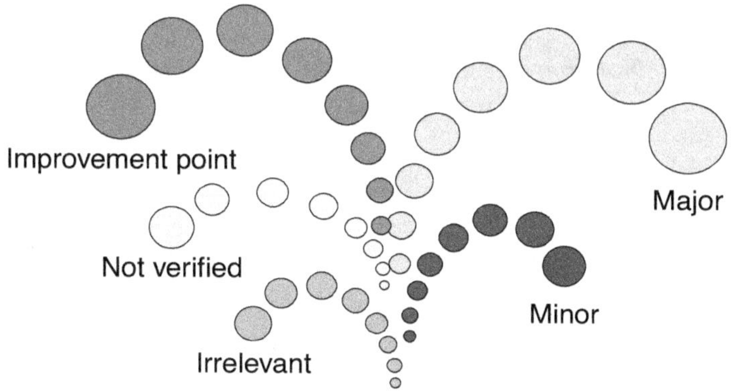

Figure 10.1 Sort your data.

Importance can be judged based on:

1. Repeated occurrences (quantitative data)
2. One time occurrences that have high risk (qualitative data)

However, observing repeat occurrences does not necessarily make the evidence important. Consequences must be considered, too (rework, loss of certification or license, nonconforming product, lost customer, patient, or client, and so on).

Qualitative data (single occurrence) may come into play regarding such issues as safety, environment, and wrongdoing (not wearing protective equipment, dumping hazardous waste, stealing, fabrication of records, and so on).

Audit evidence may be captured in any of the following types of statements:

- *Nonconformity.* Violation of a requirement that can be major or minor
- *Finding.* Systemic problem, supported by audit evidence
- *Improvement point.* An opportunity for improvement, not a violation
- *Defect.* Minor violation of little consequence
- *Concern, observation, remark, or issue.* Possible future problem for the organization
- *Positive practice, noteworthy achievement, good practice, best practice.* Some aspect of the system/process that is done very well (very effective)
- *Conformity.* Adherence to a requirement

Most auditor energy now is going into matching audit evidence with requirements (agreed-upon criteria). Reporting other data (for example, improvement points, best practices) is at the discretion of the auditor with approval from the client or audit program manager. An auditor must know the report terminology and reporting procedures prior to the audit.

Though we recommend against it, many organizations report concerns, observations, remarks, or issues.

Our concern is that the observation category becomes a dumping ground for unresolved issues, an auditor's subjective opinion, or a place for auditors to park requirements not fully investigated. Our second point is that though the auditee may say they are interested in knowing about any concerns or observations, almost nothing is ever done about them. We recommend that you continue investigating to determine if the evidence should go in a different category such as a nonconformity or improvement point. If you must, use the observation category sparingly. In this book, an observation is something the auditor does to collect audit evidence; he or she makes observations.

WRITE IT UP

You must be able to communicate the results of the investigation. One of the most common techniques is to write nonconformity or noncompliance statements. It is very important to write clear nonconformity statements so the auditee fixes the right problem and fellow auditors will be able to verify corrective actions.

When writing nonconformance statements, you may want to follow the ENRC4[1] formula: What is the *evidence* that you looked at? What was the *nature* of the nonconformity? What was the *requirement*? And, is the statement *clear, concise, complete,* and *correct* (C4)? The nonconformity statements will be the most read parts of the audit report. Writing them clearly enough allows the auditee to easily find exactly what you uncovered so that they can start correcting and investigating.

Audit Principle

Ensure that results are traceable to requirements.

Example nonconformity statement development:

- *Evidence:* Procedure 8501 incorrectly states that Purchasing handles customer complaints. It does not address the current policy that Marketing is supposed to handle customer complaints or how.
- *Nature* of the nonconformity: Documents needed to ensure effective planning, operation, and control have not been updated as necessary. When creating and updating documented information, the organization has not ensured appropriate review and approval for suitability and adequacy of use.
- *Requirement:* ISO XXX, clause 7.5.2 c).
- *Nonconformity statement* (Combining ENR): Quality management system documents have not been updated to reflect current practice. After interviews with both Purchasing and Marketing, it was observed at the audit that the corrective action procedure 8501 available for use states the Purchasing department handles customer complaints; the responsibility was changed five

months prior to the audit. 8501 does not reference that Marketing now handles customer complaints or their responsibilities. ISO XXX, clause 7.5.2 c).

Powerful nonconformity statements strive to show the opposite happening while avoiding employee names. (Steps A, B, C, and D are all required for the R55-8 process. It was found at audit that the R55-8's records for March 8, March 9, April 17, and January 3 show only steps A and D were followed.) Do not try to solve problems or point to causes or solutions in a nonconformity statement. The auditor is there to point out the nonconformity observed with the objective evidence, and then allow the auditee to resolve it together with those closer to the issue.

If you have to point to an employee, use employee titles instead of names. Management tends to blame people first but history shows us it is more likely the process, procedure, or element. Example nonconformity statements have been provided in Appendix I: Example Audit Nonconformities.

Your ability to write good nonconformity statements will improve with practice. What is important is that you communicate the problems you observed to the auditee so that they can be addressed. Many internal audit programs use some type of form to report nonconformances. It maybe a nonconformance form or a corrective action request form.

The relative importance of the nonconformities can be reported as major or minor (or using other terms such as a high versus low risk, nonconformance versus defect, or finding versus nonconformance).

Let's take a closer look at the terms major and minor for a minute.

When an external auditor would grade a nonconformity, it was in relation to how the company is following the standard. So the auditor would explain the nonconformance grading in this way:

Minor: An instance uncovered or a small trend of nonconformities found that does not conform to the standard. This is usually most of the audit findings reported.

Major: A requirement of the standard has not been implemented, or a requirement of the standard has not been addressed. Or a series of minor nonconformities found in which an auditor can deduce that a requirement of the standard has not been implemented or addressed.

In relation to ISO 9001, oftentimes an external auditor will present minor nonconformities in the exit meeting and remind the auditee that any minors not addressed by next audit will become majors—not because the auditor thinks they are special and that the auditee should do whatever the auditor says, but because the auditor found a minor nonconformity, so now the actions required from ISO 9001 should take place. When the auditor comes around next time, if no action has taken place, then it is obvious that the requirement clause was not implemented.

Some industries or company preferences have latched onto the words "major" and "minor" and interpreted the terms to be important versus not important. Some have used the words "major" and "minor" to be safety or nonconforming product going to the customer versus a regular issue, which has its merits especially for certain industries. The problem with grading in such a way in the internal audit setting is that departments will deem (and in my personal experience has actually deemed) a minor nonconformity as something that does not need to be fixed (sometimes not even addressed).

The internal audit is a time to perform an extensive look at how things are going, a health check-up of the company. These types of audits are typically deeper than any external audit. The purpose is to go deeper and find problems. So when internal auditors uncover a nonconformity they have found a problem that needs to be fixed. Everyone is on the same team internally, the same side to succeed.

Externally, an audit may have a basis to continue using someone as a supplier or to continue a company's certification or to make sure laws are being followed. External audits usually do not have an interest in your success/failure.

You may also report opportunities for improvement and best practices observed. An opportunity for improvement is an observation that is not a violation of a requirement but might improve the effectiveness of the process or organization under review. A best practice is an observation of an activity that is so outstanding that it should be shared with other parts of the organization. The subsequent implementation of a best practice by others in the organization will improve the organization's effectiveness and efficiency.

Results of an audit can also be reported as a finding. Earlier we defined a finding as a systemic problem supported by audit evidence. Audit finding statements attempt to group the causes of a problem. Most organizations report results as nonconformities due to its simplicity. Mature quality management systems may consider reporting findings to identify systemic issues. An audit finding statement is a conclusion about what was observed. Typically, there is a statement followed by the evidence supporting the finding. The supporting evidence may comprise nonconformances as well as observations related to performance issues. An example audit finding statement may look like the following:

1. The process for closure of corrective actions from audits is not effective and has resulted in repeat customer complaints, unnecessary regulatory risk, and failure to take advantage of potential cost savings.

   1.1 Six of every 10 corrective action due dates are extended at least twice.

   1.2 Three of the 10 corrective actions completed have not been reviewed to determine if the action was effective. ISO 9001: 10.2.1 d.

   1.3 Customer complaints 20416 (Acme Fitting) and 20614 (Best Buy) occurred because corrective action (CA1279) from last November has not been addressed.

1.4 Nonconformances CA1366 and CA1389 were identified as having regulatory risk (possible citation) and were overdue 35 days.

The idea behind a finding is to identify a systemic issue supported by compelling facts.

REPORT SUMMARY OR ABSTRACT

Some reports are very long and would take top management valuable time to digest and understand the stated risks and consequences. This may be particularly true of regulated organizations that report conformity as well as nonconformity. It may be difficult to page through a report to find the nonconformities. In such a case, a report summary can be used to list the nonconformities and where they are referenced in the report.

OVERALL AUDIT CONCLUSION

As lead auditor you may be asked to report an overall conclusion based on audit results. Your audit conclusion may reference a state of readiness for a pending customer audit or report the degree of compliance to internal standards (procedures and specifications) or external standards (such as ISO 9001, IATF 16949, 21 CFR 210 and 820, FAA Title 14 Chapter III, ISO 14001, ISO 13485, or OHSAS 18001). You may report any conclusion based on the evidence and your judgment or understanding of the auditee situation. From all of the evidence you gathered and nonconformance statements you generated, the overall audit conclusion can set the tone of how good/bad the audit was.

At the very minimum, an audit conclusion should be:

1. Relevant (linked to the purpose and scope)
2. Consistent with the audit evidence (based on fact)

For example, if the audit was conducted to determine the degree of compliance to ASQ/ANSI/ISO 9001, the conclusion should not be about the readiness to start up the next product line.

The conclusion should be consistent with audit evidence collected during the audit. If there were several significant nonconformities or major findings, it would not be appropriate to state that everything looked fine. If there were no nonconformities, it would not be appropriate to state that the area needs a lot of work.

The following is an example of matching the audit conclusion with the purpose:

Audit purpose	Audit conclusion
To determine the degree of compliance with ASQ/ANSI/ISO 3115 and internal department procedures.	The department is in compliance with ASQ/ANSI/ISO 3115 and internal department procedures with only a few minor nonconformances reported.

Conclusions are based on objective evidence. The auditor should point out areas of strength and weakness, because this will help auditee management decide where to concentrate their resources.

Areas of weakness or strength can be described as:

1. A standard clause, element, or control
2. An area, department, or process
3. Deployment of controls (existence of procedures and their updating)

A conclusion may also state the overall consequences *of the results* of the audit, such as:

- The area is (or is not) ready for the certification audit.
- The area is ready (or not ready) to launch the new product (service).
- There is a negligible (or significant) risk of a major regulatory citation.
- The area audit rating will increase (or decrease).
- The interval between audits will be increased (or decreased).
- A follow-up audit will (or will not) be required to continue operating.

Reporting a grade or percentile score can be considered part of an audit conclusion, such as an 'A' is an excellent rating or 77 percent matching required for ongoing approval levels. A score or grade is normally the result of some type of mathematical calculation based on the response to certain questions. Scoring provides an immediate reference to gauge an organization; however, scoring has certain shortcomings and can result in organizations implementing unneeded costly controls to achieve higher scores or resist changes to avoid risking a lower score.

Normally, the overall audit conclusions are reported in a summary, brief, synopsis section, or as pre-matter attached to the detailed audit results. The conclusion should provide the big picture (key issues of importance) for management. Some internal audits require conclusions and others do not. Good practice requires some type of conclusion because the number of nonconformities does not always reveal the true situation. An organization can have 10 nonconformities but the auditor observes a very good and solid management system. On the other hand, an organization can have four nonconformities but the auditor observes deep-rooted systemic problems that could be a risk to the organization's future.

First you will report the results at the exit meeting and then in a written report. In many cases third-party and second-party auditors also make recommendations. They may recommend certification or licensing, or recommend continued certification. They may recommend approval or disapproval of suppliers.

Now that you are organized, get ready to report your results to the auditee at the exit meeting and in written form. What do you say, and how do you say it? What if the auditee disagrees with your conclusions? Will a follow-up audit be necessary? Find out the answers in the next chapter.

ENDNOTE

1. Janet Muschlitz, *Quality Auditor Review newsletter* 1, no. 3 (1997): 4.

Chapter 11

Reporting

To finish off the audit you need to tell the auditee what you found, put it in writing, and explain subsequent (follow-up) actions.

To conclude the audit, a meeting is held with the auditee. This meeting has been called an exit, closing, or post-audit meeting.

EXIT MEETING

There must be an exit meeting to conclude the performance phase of the audit. Internal or first-party audit exit meetings are less formal than second- or third-party ones. However informal or brief the meeting, reporting results is always serious business. Exit meetings should be well organized and professional.

It is good practice to keep the auditee informed throughout the audit of any significant problem areas so that the audit conclusion will hold no surprises for the auditee. Some organizations require the auditee to acknowledge the audit findings before or during the exit meeting.

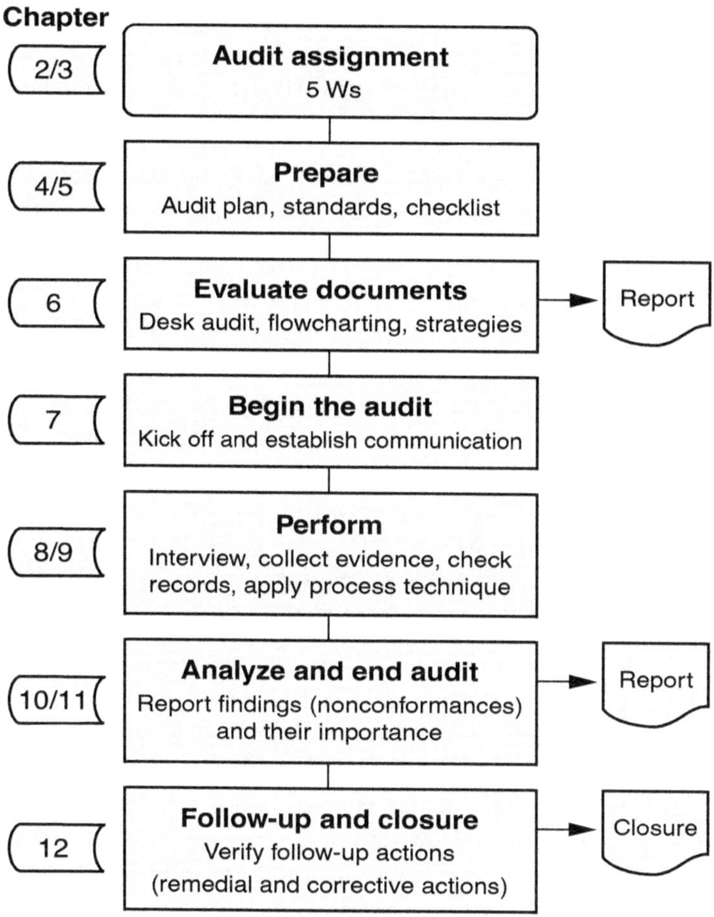

Figure 11.1 Exit meeting flowchart.

The lead auditor is responsible for the exit meeting and preparing an agenda (see Appendix C: Example

Meeting Agenda and Record). Your organization may have a set agenda based on your circumstances. Consider the following actions as you follow along the exit meeting flowchart (Figure 11.1):

- *Attendance should be taken and someone should be assigned to take minutes.* The lead auditor may assign someone to take minutes or take his/her minutes.

- *Present the purpose, scope, and method of prioritization of the results.* Inform the auditee about the classification of the observations and what it means. This agenda item may be skipped for routine internal audits. In more formal settings, sometimes this information can be taken almost verbatim from the formal opening meeting (in the opening meeting you said "This is what we intend to do" and at the exit meeting you say "This is what we did" or "as you recall from the opening meeting this was what we intended to do").

- *Distribute copies of the nonconformities (findings).* Just before you pass out the nonconformities and while you have everyone's attention, you may want to state that: "The nonconformities I am about to read may be isolated incidents requiring remedial action or they may be a symptom of a bigger systemic problem that needs corrective action and root cause analysis. It is very important to identify systemic problems to ensure that corrective action is taken and problems do not recur. Repeat problems waste organization resources and could result in additional oversight from the auditing program." We have had our share of disappointing exit meetings where management just wants to get the nonconformity checked off and out of the way without regard to the consequences to the organization as a whole. Next, pass out the findings and read the finding/nonconformity statement(s) aloud. This is a serious time. Maintain good eye contact throughout the exit meeting. Hold questions until you are finished, then ask if any of the results need to be clarified. Avoid discussing solutions, corrective action, or arguing. The nonconformities/findings are normally written on a nonconformance or corrective action request form and copies are handed over to the auditee.

By the way, the exit meeting is not the time to spring surprise nonconformities onto anyone. By the time you are reading a nonconformity, someone from the auditee should already know about it.

If an auditee objects to a nonconformance, the objection should be noted in the meeting minutes. Do not attempt to resolve the issue at the meeting. As lead auditor, you can offer to review any additional evidence after the meeting and promise to respond based on the new evidence.

- *Lead auditor presents overall conclusions.* Based on your analysis, you can present the overall conclusions. The lead auditor must present the audit findings to management in such a manner that they clearly understand the results of the audit.

- *Explain follow-up actions.* If there are nonconformities, there will be some type of follow-up to correct what was found. Follow-up action normally includes any corrective action plans. The lead auditor should also indicate any follow-up audits required as a result of the nonconformities identified. Normally, follow-up and closeout of nonconformities is handled at the next audit. If one of the nonconformities represents a high risk to the organization, a special follow-up audit can be scheduled by the audit program manager.

- *Keep records of exit meetings.* The attendance roster, results, and minutes taken during the meeting are the exit meeting records. The audit records must be safeguarded (protected). For example, ensure that extra copies of the audit report and other records are destroyed after the meeting (don't leave extra copies in the meeting room).

RESPONSIBILITIES

For the *auditee:*

- Notify personnel of the time and place of the exit meeting
- Ensure that appropriate management/ supervision is invited
- Listen to the report
- Present any additional relevant facts

For the *auditor(s):*

- Attend the closing meeting
- Support the lead auditor

- Provide clarification details if asked to do so by the lead
- Safeguard information

> **Audit Principle**
>
> *Do not disclose auditee proprietary information to others.*

PREPARE FOR THE REPORT

The report is the official product of the audit. It is the record that will be referenced when there are questions. The report must be clear and it must be written in terms the user can understand if it is to be effective. If you use a term that many may not understand, define it in the audit report.

Put the nonconformities and/or findings in order of importance (such as major and minor). Remember, your findings are only as good as the weakest one.

> **Audit Principle**
>
> *Communicate the importance of findings/nonconformities.*

REPORT FORMAT

In most cases, the audit program manager will specify a report format and provide you with report-writing guidelines. Consider the following report format points when completing the final report:

Audit Report Identification **(Title, number, other)**

Confidential classification. Company Confidential, Proprietary Information, Need-to-Know-Only Basis, Secret, and so on. Safeguard the audit report to protect its confidential nature.

Introduction or background. This section contains much of the material previously developed for the audit plan. The introduction may include: audit purpose, scope, dates of the on-site audit, standards audited against, auditee organization and areas audited, client, the auditing organization, and the audit team members.

Qualification/limitations. Report any sampling limitations or scope changes. Reflect on issues that may qualify the results, such as: the company may produce all kinds of brackets, but only "X" brackets were being fabricated during the audit.

Conclusion/summary. Overall assessment as to conformance to the standard or achievement of the management system objectives.

Best practice/noteworthy achievement. Report the good things found during the audit.

Detailed audit results. Details of the major/minor nonconformities/findings.

Improvement points. Report if agreed upon prior to the audit.

Report by (your signature) and date.

Audit Principle

Report the results of the investigation truthfully and in a clear, correct, concise, and complete manner.

Turn in your report as required. Many internal audit programs require the auditor to submit the audit report to the audit program manager for approval and distribution. In other cases, the report automatically goes to the area audited with copies going to the audit program manager.

WHAT TO AVOID

- Using emotional words and phrases such as: "grossly mismanaged," "totally out of compliance," "there is absolutely no management commitment," and so on. Such statements will get management attention but are unlikely to lead to improvement.

- Using words that may create the appearance of bias or a slanted viewpoint.
- Reporting minor imperfections found during the audit if there is no potential added value from their correction. One of the Four Audit Management Realities[1] is that "nothing is perfect." As an auditor, you can always find something wrong. Looking for imperfection is more akin to inspecting, not auditing.
- Reporting names of individuals unless it is germane to understanding or correcting the problem found.
- Making recommendations or telling auditee how to go about addressing the nonconformity.

RECOMMENDING SOLUTIONS

Good audit practice is that auditors should not take ownership of the problems identified during the audit. Making recommendations implies that the auditor has the ready-made solution for the problem or nonconformity.

Making recommendations can result in the following outcomes:

- Auditee implements the recommendation even though they may know it is wrong just to get the report closed out. This is called *malicious compliance* by the auditee.

- Recommendations are ridiculed as being unrealistic or even silly due to the auditor's lack of process knowledge of the area audited.
- The auditee becomes defensive and will not recognize or affirm even a good recommendation. The auditee may actually implement a suboptimal solution just to avoid lending any credence to the auditor's recommendation.
- When the auditee expects the auditor to come up with solutions to any problems, there will be an auditor bias to find fewer problems.
- If asked to audit the same area later, the auditor's objectivity would be compromised.

Audit Principle

Do not take ownership of problems found

When audit program management requires auditors to make recommendations for corrective action of audit nonconformities, the auditor must comply. A technique for helping but not telling auditees how to fix a problem is to provide examples of how others have addressed similar problems. Also, the auditor making recommendations should not audit the area again to verify the corrective action.

In order to take full advantage of the knowledge and skills of the internal auditing team, some organizations assign auditors as advisors for areas they will never audit. The area personnel can ask their advisor for input in taking corrective action.

ENDNOTE

1. J.P. Russell and T. Regel, *After the Quality Audit* (Milwaukee: ASQ Quality Press, 2000).

Chapter 12

Audit Follow-Up, Corrective Action, and Closure

The auditee is responsible for fixing what was found during the audit, and the client is responsible for following up and determining the extent of the auditor's involvement in follow-up actions. Normally, an auditor is assigned to follow up on actions taken to address audit findings.

The determination of who is responsible for following up audit findings may be a function of the business, organization culture, liability, risk, and/or the availability of competent resources. Regardless of who is assigned follow-up responsibility, auditors should be aware of the corrective action process and proper follow-up steps to ensure that problems are fixed.

ELEMENTS OF THE CORRECTIVE ACTION (CA OR SOMETIMES EVEN CAPA) PROCESS

Let us assume that an audit report has been issued and there are nonconformities that require corrective action. The auditee has agreed to submit a corrective action plan to the audit organization by an agreed-upon date. Now, as the lead auditor, audit program manager, or client, you must review the corrective action plan submitted by the auditee. It is the auditee's responsibility to take corrective action and issue the corrective action plan.

The corrective action plan should be issued within a specified time agreed upon between the audit organization and the auditee. If the corrective action is not on time, it is overdue.

The corrective action plan should contain the following:

- Definition of the problem or restatement of the finding.
- Remedial action (containment, correction, countermeasures). This is considered temporary for a finding requiring corrective action. Or for very minor issues it can be the stopping point if no corrective action is necessary.
- Measurement and data gathering: What is the root cause and how can it be eliminated?
- Solution(s): Eliminate the cause of the problem, nonconformity, or undesirable situation.
- Measures to determine whether the corrective action was effective.
- Action plan steps (the Do, Check, Analyze steps).
- Responsibilities and due dates.

See Appendix K: Example Corrective Action Request.

The auditee proposes the solution and determines the importance of fixing the problem. Too often auditees want the auditor to tell them what to do to close out the finding, but that is not considered good practice. It is important for the auditee to assess the importance of the finding and respond (act) accordingly (work on the important stuff). It is perfectly okay to take remedial (containment) action as a first step toward corrective action or to address minor nonconformities that do not represent a systemic problem.

Remedial actions (containment, correction, countermeasures, quick fixes) only address the immediate nonconformity or defect. They include reworking, rejecting, repairing, re-grading, replacing, releasing as-is, or retraining. Remedial actions do not eliminate the cause of the non-conformity. The nonconformity will recur unless it is an isolated incident and not likely to ever happen again.

Please note that ISO 9000 uses the term *correction* to describe repair and rework activities. However, the nuance between making corrections and taking corrective action is confusing. It would be best to use the terms *remedial action* and *corrective action* where applicable.

The corrective action plan is submitted for review and approval. You may not be the one reviewing the corrective action plan but later on you may verify actions taken and their effectiveness. The reviewer should determine if the root cause has been identified and the stated corrective action plan is consistent with the stated finding. The review output may be a simple matter of acknowledgment of the action to be taken.

The reviewer verifies that the actions address issues relevant to the finding and that they are adequate to provide a complete solution. A corrective action plan may be rejected because (1) the finding is not addressed, (2) the root cause is not identified, (3) priority or timing is not appropriate, or (4) relevant information is missing.

The auditee may claim that no corrective action is necessary and provide additional information to support their claim. In other cases, the auditee may request more time to address the finding. Normally, a request for an extension should be granted unless it is a safety or environmental (high-risk) issue or the courtesy of granting extensions has been abused.

VERIFICATION METHODS

Corrective actions should be verified according to established procedures and methods (step 4 in Figure 12.1). Methods for corrective action verification include:

- Verification during a subsequent audit of the same area (same or different auditor)
- Scheduling a follow-up audit specifically to verify the corrective action(s) (same or different auditor)
- Examination of implementation and performance records provided by the auditee

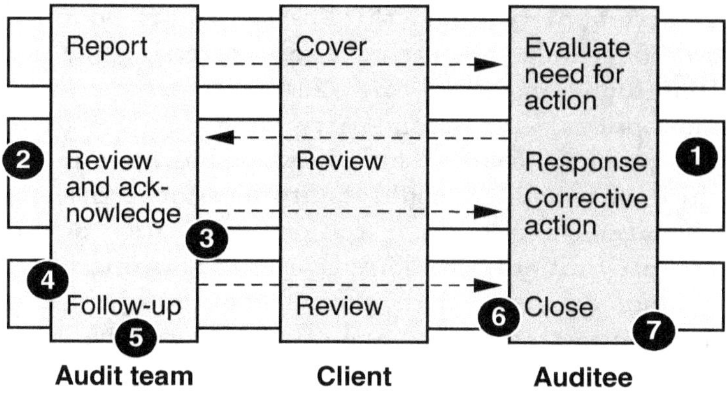

Figure 12.1 Audit follow-up cycle.

Corrective actions can be verified one at a time regardless of the source or number of corrective action requests from a single audit. Corrective actions can be tracked and closed individually.

THE FOLLOW-UP AUDIT

The client will determine if a follow-up audit is required (step 5 of the audit follow-up cycle). If a follow-up audit is required to verify that the corrective action has taken place, it should be scheduled to allow sufficient time for implementation. The auditee should be notified of the follow-up audit, and standard audit conventions should be practiced. The follow-up audit can be conducted by the same or different auditor(s). See Appendix L: Corrective Action Checklist for a checklist for verification of the implementation of the corrective action.

Second-party and third-party audits normally are done under a contract. Thus, correcting the problems found in second-party and third-party audits is not optional. For second-party audits, failure to correct problems could result in loss of business, and for third-party audits it could result in loss of certification (management system registration/certification or product certification) or endorsement of the organization or product. Because of the commitment of the organization (the contract), follow-up and effective corrective action becomes a very serious matter.

The completion of the corrective action plan and its implementation should be verified. The investigation can include verification of document changes or employee awareness of the changes, observation of work practices, and review of records. There should be a record confirming that the corrective action completion was verified. An example would be signing or initialing and dating a section of a corrective action form or report (or both).

EFFECTIVE CORRECTIVE ACTIONS

Besides verification that the corrective actions were implemented, auditors or other assigned persons should verify that the corrective actions were effective. The auditee should be required to list the measures taken to determine if the corrective action was effective.

There are two elements involved in determining if the corrective action was effective:

1. Did it achieve the desired result? This is proof that the process improved and the actions implemented are consistent with business goals.
2. Is the process capable and efficient? There is evidence that the process will consistently achieve the desired result in a cost-effective manner.

CLOSURE

Action has been taken on the audit finding, has been implemented, and is effective. All that remains is closing out the finding (corrective action request). The closeout should include:

1. A record of the closure (letter, memo, report, or meeting minutes)
2. Communication of the closure information to the client (and in turn to the auditee)
3. The time the corrective action was completed compared to what was originally promised (actual versus original estimate)

In some cases with internal audits, the closure is the signature of the auditor and audit manager with the filing date or the date it is sent to top management. In most cases the closure notification is sent to those on the original report distribution list. Others who might receive a copy of the closure notification from the client are top management, the department managers, and the audit program manager. It is also okay to hold a meeting with the auditee to formally close the corrective action and note the success in doing so. This meeting has been called an exit, closing, or post-audit meeting.

Upon audit closure, it is a good time to discard all working papers associated with the audit except for the formal documents and records. Make sure this is appropriate first, as some organizations want to retain everything. The working papers should not be needed because the real star is the completed audit report; it should have everything that is needed for future reading or reference.

Timeliness with regard to corrective action implementation is not considered to be a fixed time period (such as 30 days) but a specified time period. The specified time should be based on the importance (effect on the organization) of the corrective action and the availability of resources. Important issues should not be held up for 30 days and complicated problems should not be rushed. Corrective actions completed on schedule should be considered timely. Schedule delays could be due to a lack of resources. The auditee should keep the auditing organization informed of any delays and the reason for the rescheduled implementation. The auditing organization (or other designated function) should monitor auditee progress.

When done right, auditing provides valuable information to management concerning compliance and performance of areas under management control. When done wrong, auditing creates conflict, blocks the achievement of objectives, and wastes organization resources.

I want to thank you for choosing Internal Auditing Fundamentals. I want to wish you success and hope you enjoy auditing as much as I do.

—Paul

Appendix A
Example Audit Plan

AUDIT PLAN—9/10/20XX

Audited Area

Quality/Environmental Management Assurance and Operations

Purpose

Verify conformance (compliance) of the system using ASQ/ANSI/ISO 9001 quality system standard and department operating procedures and to report any nonconformance(s).

Scope of the Audit

The quality and operation activities relative to improvement and the corrective and preventive action (CAPA) plans. The audit includes all inputs to corrective action and follow-up actions.

Requirements

As specified in ASQ/ANSI/ISO 9001 Clauses 10.1, 10.2, 10.3, HACCP, and existing department policies and procedures.

Applicable Documents

The quality manual sections:

8503 Continual Improvement

8502 Corrective Action Procedures and CAPA

Other regulatory requirements and industry standards

Schedule

November 6, 20XX

8:15 am . . . . . . Opening meeting with area manager

8:30 am . . . . . Interviews

November 7, 20XX

6:00 am . . . . . Night shift interviews

9:00 am . . . . . Exit meeting

Team Members

Paul Horseshoe

Mark Anvil

Approved: ______________________________

Paul Horseshoe, Audit Team Leader, ASQ CQA

Approved: ______________________________

Jane Smith, Audit Program Manager

Appendix B

Example Work Order

To: Distribution
From: A. J. Auditor
Date of WO: September 12, 20XX Work Order Number: X1106QA

Audit Date(s): November 6, 20XX
Auditor(s): A. J. Auditor, Mary Goforth, Linda Hopewell

Client: Audit program manager
Contact: Bill Sims, improvement and productivity manager
Phone: 888.888.8888 x 123 Fax: 999.999.9999

Area/location: Anytown
Standard: ASQ/ANSI/ISO 9001, clause 10 Improvement and applicable QMS procedures.
Purpose: To determine the Quality Assurance and Operations departments' adherence to continual improvement and corrective action requirements and to verify closeout of prior nonconformances.
Scope: ABC Company QA and Operations activities regarding corrective action.

Audit Services:

___ Lead auditor
_X_ System audit
___ Process audit
___ Product audit
_X_ Verify corrective actions from prior audits
___ Desk audit
___ Gap analysis
___ Report opportunity for improvement
___ Report recommendations
___ Qualify/training new auditors
___ Combined audit
___ Audit multiple shift operations

___ Other: ______________________________

Preliminary Audit Schedule:
November 6, 20XX
8:15 AM Opening meeting
8:30–4:00 PM Interviews

November 7, 20XX
Exit meeting

Action Item List:
AJA—Conduct desk audit by 10/24
ADM—Get copies of corrective actions from prior audit and send to audit team

Comments: Need to discuss possible interviews 6:00 AM on November 7 for overnight shift prior to exit meeting.

Distribution: Immediate supervisor, audit program manager

Appendix C

Example Meeting Agenda and Record

Audit Meeting Record

Meeting2

Meeting date: ____________________ Time: ____________

Location: ____________________ Audit no.: ____________

Meeting Agenda

Opening meeting Date and time: ____________

- ❑ Introductions: lead auditor, auditors, observers, auditee representatives; take attendance
- ❑ Thank-you
- ❑ Audit scope: location, areas, units, department
  Verify access to areas to be audited
  Identify safety and environmental requirements or issues
  Know emergency evacuation routes and protective equipment required
  Security and confidentiality issues
- ❑ Audit purpose: verify conformance to ISO 9001 (or other standard or procedure) and follow up prior audits
  Other includes: training, evaluate the audit process, communicate to management
- ❑ Methods and techniques: conduct interviews, examine records, make observations
  Sample selection, tracing techniques
  Report results as a major/minor nonconformity or improvement points
- ❑ Detailed audit schedule: pass out or display interview schedule and resolve conflicts (persons not available, lunch, quitting time)
- ❑ Confirm logistics (audit team area/desk, rest rooms)
- ❑ Confirm exit meeting time and attendance

Exit Meeting Date and time: ____________

- ❑ Thank-yous for hospitality and cooperation. Take attendance.
- ❑ Review audit purpose and scope (audit plan) and explain how results will be reported
- ❑ Pass out detailed findings and present (nonconformity, conformity, improvement points)
- ❑ Positive practices
- ❑ Audit conclusions
- ❑ Follow-up actions
- ❑ Other: ____________________

Meeting Minutes

____

____

____

____

Meeting Attendance Roster

Actual time: __________ Actual date: __________ Audit no.: ____________

Print name	Department/function	Opening	Exit

Appendix D
Example Interview Schedule

Table D.1 Detailed audit schedule for home office audit—Issue 3.

Assignment	Area/function	Contact	Auditor	Date/time
Day one				**Oct. 27th**
Auditor team meeting	2nd floor conference room		All	8–9:30 AM
Opening meeting	2nd floor conference room	Selected by management representative	All	9:30–10 AM
8.4.1 Purchasing and QMS 7401		Dennis Power, manager	JR	10–10:30 AM
8.4.1, 84.2, and QMS 7401		Daniele Cable, buyer	JR	10:30–12 noon

continued

continued

Assignment	Area/function	Contact	Auditor	Date/time
Day one				**Oct. 27th**
7.5, and QA 4232		Department coordinator	JR	1–2 PM
8.4.3 and QMS 7405	Issuing POs	Bob Port, clerk Dave Beam	MH	10 AM–12 noon
8.4.2 and QA 7439	Receiving and inspection	John Transom	MH	1–2 PM
Prepare report	Conference room		All	2:00–3:00 PM
Exit meeting	Conference room	Auditee personnel	All	3:00 PM

Appendix E
Example Checklist Page

Standard: ASQ/ANSI/ISO 9001	Audit number: ______________		
Ref.	**Question** *[Italic statements are not in the standard]*	**Yes/ no**	**Comments** (Evidence—data)
8.2	**Requirements for products and services**		
8.2.2	**Determining the requirements for products and services (or pre-contract)**		
8.2.2-1	*Can the organization meet its claims for the products and services it offers?*		
8.2.2-2	Does the organizations ensure that the requirements for the products and services are defined, including obligations such as regulatory and legal requirements, and additional requirements considered necessary by the organization? *[This is a prescriptive list. Verify that items are addressed such as under review of customer requirements. For example, there could be a nonconformity for not defining necessary but unspecified customer requirements.]*		
8.2.3	**Review of requirements for products and services (or contract)**		
8.2.3.1-1	Are specified customer requirements (new or changed contracts, tenders, and orders) including requirements for delivery and post delivery activities reviewed prior to commitment?		
8.2.3.1-2	Are requirements that are necessary for the specified or intended use, even those not stated by the customer, reviewed?		
8.2.3.1-3	When requirements are not written (documented by the customer), are they confirmed by the organization before acceptance?		
8.2.3.1-4	Are contracts or order requirements that differ from those previously expressed (in the tender or offer) resolved?		
8.2.3.1-5	Are customer requirements reviewed to ensure that the organization has the ability to meet them?		
8.2.3.16	Does the review include requirments specified by the organization?		
8.2.3.1-7	Does the review include statutory and regulatory requirements applicable to the products and serviices?		
8.2.3.2	**Is there documented information available as evidence regarding . . . .**		

Checklist dated: April 10, 20XX.

Appendix F
Audit Time Considerations

Size of organization: need to take reasonable sample (interviews)

Complexity of operations (number of different products/services or different activities): several different types of operations, activities, products need to be sampled

Technical level

Number of requirements to be assessed against: size of performance standard and level of difficulty

Meeting times: opening, closing, team, briefings

Logistics: lunch, breaks, distance from area to area

Remote audit considerations: factoring in the resources needed to conduct successful remote portions along with the time of those involved in the audit preparation and audit performance.

Special training: safety orientation, environmental

Tour or time to better understand operation

Report preparation time

Contingency time

Work hours of auditee

Shift operations: number of shifts to be observed

Organizational culture: more formal cultures may require longer meetings and formal requests for records

Past audit performance data can also be utilized for estimates of future audits in the same areas/locations.

Security

Appendix G
Example Notification Letter

Jeb Blacksmith October 14, 20XX Go Industries
Department 3221
1234 Mile Road Your Town, PA 18111

Dear Jeb,

We plan on performing a quality system audit of the Quality Department and Operations to assess conformance to ASQ/ANSI/ISO 9001:20XX standard requirements and area procedures. We will limit the scope of the audit to departmental activities concerning corrective action and Improvement. During the audit we also plan to verify the corrective action of past nonconformities. The audit process will be conducted according to company audit guidelines and practices.

We plan to arrive at 8:00 am and hold the opening meeting at 8:30 am. We plan on auditing the afternoon shift between 4 and 6 pm. Due to the late hour and possible interviews of the overnight shift, we plan for the exit meeting to take place the next day. I will leave a draft report, with the final report following within five business days. The auditors will be myself and Mark Anvil.

The on-site audit is scheduled for November 6. The audit plan is enclosed. Please provide a production schedule, orders-shipped list, and list of equipment under calibration control NLT November 1.

Thank you for your cooperation. If you have any questions please contact me by telephone or e-mail.

Sincerely,

Paul Horseshoe

Copy: Charles Bellows, client
Mark Anvil

Enclosures: 1

Appendix H

Popular Performance Standards

ASQ/ANSI/ISO 9001, *Quality management systems-Requirements.* Used by organizations to establish a fundamental quality management system and for registration/certification purposes.

ASQ/ANSI/ISO Q9004, *Quality management systems-Guidelines for performance improvements.* Used by organizations to advance their quality management system to improve its effectiveness and efficiency.

ISO/TS 16949, *Quality systems-Automotive suppliers-Particular requirements for the application of ISO 9001.* A technical specification used to meet fundamental auto industry sector requirements.

SAE AS9100 The International Aerospace Quality System Standard. Used by organizations in the aerospace industry to establish a fundamental quality management system and for registration/certification purposes.

ASQ/ANSI/ISO E14001-2004, *Environmental management systems-Requirements with guidance for use.* Used by organizations to establish a fundamental environmental management system and for registration/certification purposes.

ASQ/ANSI/ISO QE19011S-2004, *Guidelines for quality and/or environmental management systems auditing.* Used by auditors and auditing organizations to establish a fundamental audit program for internal and external audits. Replaced ASQ/ANSI/ISO 10011, ANSI/ISO 14010, ANSI/ISO 14011, and ANSI/ISO 14012.

BS 8800:2004, *Guide to occupational health and safety management systems.* A British standard that addresses safety needs.

Department of Health, Education, and Welfare, FDA, Good Manufacturing Practices for Food, Drug, Medical Devices, and Cosmetics, Part 820—Quality System Regulation (effective 6/1/97).

ISO 13485, *Medical devices-Quality management systems-Requirements for regulatory purposes.*

OSHAS 18001 Health and Safety Zone. An international occupation health and safety management system specification comprising 18001 and 18002.

Appendix I

Example Audit Nonconformities

YOUR COMPANY AUDIT REPORT FINAL

Definitions Used in the Audit Report

nonconformity—The nonfulfillment of requirements. Specified requirements include the ASQ/ANSI/ISO 9000 series quality system standards and the existing quality management system.

major nonconformity—The nonfulfillment of ISO 9XXX–specified requirements that would prevent registration of the facility.

improvement point— Improvement points are not nonconformities but are observations that point out possible inefficiencies and opportunities for improvement.

Examples

1. *Nonconformity.* YourCompany claims application of controls for 'customer property' and 'post-delivery processes,' but such controls are not used or necessary due to the nature of the business. ISO 9001 clause 4.3.

2. *Nonconformity.* When changes were made, it is not apparent they were carried out in a planned manner. No actions were considered to ensure the integrity of the existing QMS. ISO 9001, clause 6.3 b.

3. *Nonconformity.* Seven of twelve personnel were not aware of the quality objectives and how progress (effectiveness) is reported (weekly 200X Quality Goal progress sheets). ISO 9001, clause 6.2.1 f.

4. *Major nonconformity.* At audit there was no record that employees are competent (trained or knowledgeable in the required job duties) to do the assigned job (for promotions and job transfers). ISO 9001 clauses 7.2 d and 7.2 a.

5. *Major nonconformity.* There is no evidence that training and other actions to assure competence are evaluated for effectiveness. This is the case for promotions and job transfers. ISO 9001 clause 7.2 c.

Note: I did not determine if effectiveness of new employee training was evaluated.

6. *Nonconformity.* A significant number of personnel working the day shift on June 8 were not able to express how they contribute (relevance and importance) to meeting the quality objectives (customer requirements). ISO 9001 clause 7.3 c.
7. *Nonconformity.* Design outputs, such as on Form R&E-B1.0, are not approved prior to release. Approval is not required in the procedure and there is no space on the form for approvals. ISO 9001, clause 8.3.5.
8. *Nonconformity.* Management and supervision in shipping and storage could not demonstrate how the conformity of the product is ensured. Damage is noted and issues are discussed with area personnel; however, there is no metric to monitor whether actions are effective (not able to link management controls to results). There is no metric demonstrating that the conformity (damage free condition) of the product is being maintained. ISO 9001, clause 8.5.4.
9. *Nonconformity.* There is no record of the potential effect and possible consequence of nonconforming product that has already been delivered. ISO 9001, clause 8.7.2

Appendix J

Auditor Code of Conduct*

1. I will be honest and impartial.
2. I will hold paramount the safety, health, and welfare of the public in the performance of my duties.
3. I will perform my duties in a professional manner by following procedures and doing what is reasonable and normally expected.
4. I will perform services in areas where I am competent.
5. I will not represent conflicting or competing interests and will disclose to any client or employer any relationships that may influence my judgment.
6. I will not accept any inducement, commission, gift, or any other benefit from the auditee or competing organization.
7. I will not intentionally communicate false or misleading information.

---

* Adapted from ISO 19011 principles, ASQ Code of Ethics, and Code of Conduct for RABQSA Certified Auditors.

Name: ________________________________

Signature: ____________________________

Date: ________________________________

Appendix K

Example Corrective Action Request

Corrective Action Request—page 1

Date:		Number:
Improvement ❑	Audit ❑	Audit no.:
Area:		
(A) Finding/problem		

Contact	Auditor/originator
Signature: ______________________	Signature: ______________________
Printed name:	Printed name:
Telephone number:	Telephone number:

(B) Remedial action:

(C) Root cause:

(D) Action plan:

Measures:

	Start	Complete	Acceptance	
Corrective action plan dates:			Manager/ auditor:	Date:

Corrective Action Request—page 2		
		Number:
(E) Corrective/preventive action taken: *Select alternatives—determine measures—implement—evaluate results*		
(F) Effect/explain: Reduce cost ❑ Opportunity ❑ Avoid risk ❑		
(G) Recommended actions for other areas:		
Team leader signature:		Date:
Corrective action approved ❑ Comments:	Corrective action disapproved ❑	Manager/ auditor:
Follow-up (audit) date:		Signature:
Closeout date:		Signature:

Appendix L
Corrective Action Checklist

The following checklists can be used to verify that corrective action was effectively implemented and that it is effective.

Effectively Implemented (Deployed)?*

1. Were operating personnel made aware of the change and the purpose of the change?
2. Did the organization consider what would be required to implement the change such as classroom training, on-the-job training, examples, new standards, directives issued, operator surveys, and so on?
3. Were all relevant documents, system requirements, and record-keeping requirements modified to reflect the change to the process/system?
   - Instructions?
   - Bills of materials?
   - Formulations?

---

* Taken from J.P. Russell and T. Regal, *After the Quality Audit* (Milwaukee: ASQ Quality Press, 2000).

- Testing and inspection?
- Specification and design?
- Packaging and markings?
- New purchased materials?
- Product storage?
- Service procedures?

4. Is the person or function responsible for the authorization of process changes clearly designated?
5. If the product or service provided to the customer is affected, was the customer notified of the change?
6. Is the change being followed consistently?

The following checklist can be used to verify that the corrective action was effective.

Effective Corrective/Preventive Action?

1. Were output measures identified to monitor and verify that the process is achieving the desired result?
2. Have desired expectations (outputs) been defined? Are outputs consistent with expectations?
3. Were process measures identified to monitor and verify that the process/system is capable?

4. Have desired expectations (processes) been defined? Are process measures consistent with expectations?
5. Are there records of the results?
6. Is an emergency change procedure in place to prevent nonconforming product or service if the change is not working?

Appendix M

20 Basic Audit Principles

AUDITOR CONDUCT

1. Do not disclose auditee proprietary information to others.
2. Be honest and impartial by avoiding conflicts of interest.
3. When an unethical activity is observed, verify it, record it, and report it.
4. Protect auditee property entrusted to you.
5. Use knowledge and skills for the advancement of public welfare.

PREPARING

6. Ensure that sufficient resources are available to accomplish the purpose of the audit.
7. Verify that there is an established system/process to audit before the audit.
8. Assigned auditors must be competent/qualified.

9. Communicate agreed-upon information to the auditee, such as audit times, purpose, areas to be audited, and standards to be audited against.

PERFORMING

10. Verify conformance to agreed-upon requirements (the rules). Auditors don't determine auditee requirements.
11. Ensure that sufficient samples (records, product, processes, interviews, and so on) are taken to match the purpose and scope of the audit.
12. Stay within the agreed-upon scope unless the degree of risk necessitates other actions.
13. Samples must be random and representative unless specified objectives require otherwise.
14. Conformance and nonconformance must be verifiable and traceable.
15. Comply with auditee rules (safety, environmental, health, restricted areas, and so on).
16. Keep auditee informed of audit progress.

REPORTING

17. Report the results of the investigation truthfully and in a clear, correct, concise, and complete manner.

18. Communicate the importance of findings/nonconformities.

19. Ensure that results are traceable to requirements.

20. Do not take ownership of problems found.

Glossary

acceptance criteria—Predetermined desirable characteristics that will meet customer requirements.

attribute data—1) A quality characteristic classified as either conforming or nonconforming to specifications. 2) Data requiring a count of discrete measurements such as good and bad, used when variable measurements are not possible (color, missing parts, scratches, damage, smoothness) or where go/no-go gauges are preferred over taking actual measurements (hole diameter range, over/under, align with template).

audit—1) Systematic, independent, and documented process for obtaining evidence and evaluating it objectively to determine the extent to which audit criteria are fulfilled. 2) A planned, independent, and documented assessment to determine whether agreed-upon requirements are being met. Ref. ASQ Quality Auditing Technical Committee (now the Quality Audit Division of the American Society for Quality).

See quality audit.

audit evidence—Records, statements of fact, or other information that are relevant to the audit criteria and are verifiable. Note: "verifiable" in the sense that they can be cross-checked.

audit plan—Description of the on-site activities and arrangements for an audit. Simply, it is a plan for the audit that can take on any form convenient for the auditors and auditee.

auditee—Organization being audited.

auditor—1) Person qualified to perform audits. 2) A person with the competence to conduct an audit.

best practice—Something observed that is outstanding and should be shared. Some- times called "noteworthy achievement" or "positive practice."

client, audit—The organization or person requesting the audit.

competent—1) Having requisite or adequate ability or qualities. 2) Having the capacity to function or respond in a particular way. Competence denotes having acquired and to be using one's formal education, training, skills, and experience. 3) Demonstrated ability to apply knowledge and skills.

concern, audit—Issues that are potential nonconformities.

concession—Permission to use or release a product that does not conform to specified requirements. Note: a concession is generally limited to the delivery of a product that has nonconforming characteristics within specified limits for an agreed time or quantity of that product (ISO 9000, 3.6.11).

conduct—A mode or standard of personal behavior especially as based on moral principles.

conformity—Fulfillment of a requirement.

conformity assessment—Conformity assessment includes all activities concerned with determining directly or indirectly that relevant requirements in standards or regulations are fulfilled [NIST].

continual improvement—A process of ongoing changes that add value to an organization. Also known as continuous improvement. Continual improvement is thought (by some regulators) to be a step-wise improvement, as opposed to continuous improvement, which is thought to be perpetual or constant. Continual improvement is a recurring process of enhancing the environmental management system to achieve improvements in overall environmental performance consistent with the organization's environmental policy.

continuous improvement—Includes action taken throughout an organization to increase the effectiveness and efficiency of activities and processes to provide added benefits to the customer and organization. It is considered a subset of total quality management and operates according to the premise that organizations can always make improvements. Continuous improvement can also be equated with reducing process variation.

control—1) Power or authority to guide or manage, directing or restraining domination. 2) "Effective control" is when management directs events in such a manner as to provide assurance that the organization's objectives and goals will be achieved [Statement from Internal Auditing Standards Glossary]. 3) Control is when the requirements of clause 7.5.1 of ISO 9001 have been implemented and maintained.

control plan—Documented descriptions of the systems for controlling parts and processes to provide control of all characteristics important for quality and engineering requirements. There is also a similar document called a quality plan that includes control of projects, products, processes, or contracts. ISO 10005, *Quality management-Guidelines for quality plans* has more information.

correction—Action taken to eliminate a detected nonconformity. Correction may involve repair, rework, or regrading.

corrective action—1) Action taken to eliminate the causes of an "existing" nonconformity, defect, or other undesirable situation in order to prevent "recurrence" (reactive). 2) Action taken to eliminate the cause of a detected nonconformity or other undesirable situation.

corroborate—1) Confirm, verify, authenticate. 2) To support with evidence or authority, to make certain.

credibility—1) The quality or power of inspiring belief. 2) Capacity for belief. Note: "credible" is defined as offering reasonable grounds for being believed.

customer—Organization or person that receives a product.

customer property—Property provided by the customer and owned by the customer. This can include raw materials, packaging, methods, and intellectual property.

defect—Nonfulfillment of an intended usage requirement or "reasonable expectation," including one concerned with safety.

directed sampling—Directed (or judgmental) sample selection is based on the auditor's judgment or direction given to the auditor. The auditor may purposely bias the sample selection to only high-risk or problem areas.

discovery sampling—A random sampling technique that uses no methodology. Easy to use but could result in biased samples.

effectiveness—1) Extent to which planned activities are realized and planned results achieved. 2) The consideration or balance between achieving the desired results (the product) and how they were achieved (the process). 3) The degree to which objectives are achieved in an efficient and economical manner.

efficiency—1) Relationship between the result achieved and resources used. 2) Accomplishes objectives and goal with optimal use of resources.

environment—Surroundings in which an organization operates, including air, water, land, natural resources, flora, fauna, and humans, and their interrelations.

ethical—1) Of or relating to the field of ethics or morality. 2) Involving or expressing moral approval or disapproval. 3) Conforming to professionally endorsed principles and practices.

ethics—1) The discipline dealing with what is good and bad or right and wrong or with moral duty and obligation. 2) A—a set of moral principles or values; B—a theory or system of moral values; C—the principles of conduct governing an individual or a group.

evidence—Data (records, responses to questions, observations, and so on) that can be verified. Also called "objective evidence." Evidence can be qualitative and/or quantitative. *See* audit evidence.

finding—1) Deficiency found during an audit. 2) The result of an investigation. 3) A type of audit result that makes a statement about systemic problems. 4) Results of the evaluation of the collected audit evidence against audit criteria.

flowchart—A picture of the separate steps of a process in sequential order. Sometimes called a process flow diagram or service map.

gig list—A list of minor infractions.

haphazard sampling—Selecting a sample with a goal to be as random as practical and representative of the population being examined.

improve—To enhance in value or quality: make more profitable, excellent, or desirable.

improvement point—Areas of ineffectiveness or poor process efficiency.

information—1) Meaningful data. Examples are records, procedures, and work instructions in any medium. 2) Something received or obtained through informing, such as knowledge communicated by others or obtained from investigation, study or instruction.

inspection—Activities such as measuring, examining, and testing of characteristics against predetermined acceptance criteria to determine conformity.

method—1) A plan or system of action, inquiry, analysis, and so on. 2) Order or system of one's actions. 3) The manner in which one acts, as in conducting business. Note: methodologies may be a body of methods, rules, and postulates employed by a science, art, or discipline.

noncompliance—Term used in place of *nonconformity*; popular in the regulated industries.

nonconformity—Nonfulfillment of a specified requirement, or nonfulfillment of a requirement.

Grading nonconformities, if needed:

Major (a varied definition, entirely dependent on industry or company preference): Usually it is a clause in a standard that was never implemented or never addressed after implementation,or a large trend of nonconformities uncovered that a reasonable person can see that a clause was not addressed or implemented; it can also mean that no documented information was found regarding a requirement; it can also mean a safety or environmental concern within scope uncovered at audit; it can mean that nonconforming product was released to customer impacting the customer; it can mean a law broken versus an office policy not followed; it can even mean an issue that should be fixed within 30 days. As you can see, defining "major" in an audit report is dependent on the industry and organization. Some companies will not grade findings for internal audits because all findings should be addressed quickly, just be aware and be consistent with expectations.

Minor (a varied definition, entirely dependent on industry or company preference): Usually it is an instance or a small trend of instances, or an emerging trend of instances where a requirement or policy is not being met; it can also mean an office policy not followed versus a law or regulation being broken; it can also mean there are minor inconsistencies between procedures and how they are being followed; it can even mean an issue that should be fixed within 90 days. As you can see, defining "minor" in an audit report is also dependent on the industry and organization. So, again, just be consistent with expectations.

objective—A) Uninfluenced by emotion, surmise, or personal prejudice. B) Based on observable phenomena, presented factually.

objective evidence—Data supporting the existence or verification of something.

observation—Something viewed. During an audit or investigation, an observation could be information that may be evidence to support audit findings.

organization—Group of people and facilities with an arrangement of responsibilities, authorities, and relationships. Note: where "supplier" was used in the 1994 version of the ISO standard, "organization" is now used.

PDCA—The plan–do–check–act (PDCA) cycle was first developed by Shewhart and then popularized by Deming.

planned arrangement—A planned arrangement could be any predetermined method such as a procedure, outline, checklist, or other means.

prescriptive—Requirements that are very specific and detailed. These types of requirements are not subject to wide interpretation.

procedure—1) A document that provides information for carrying out a process or activity in an orderly manner (the document can be in any medium). 2) A document that specifies a way to carry out an activity. 3) A set of steps that should be followed when seeking a desired effect.

process—1) A set of interrelated or interacting activities that transforms inputs into outputs. 2) A series of steps leading to a desired result. 3) A set or series of conditions, operations, or steps working together to produce a desired result.

process audit—1) An audit of the elements (conditions and resources) supporting an activity or process. 2) An analysis of a process and appraisal of the completeness and correctness of conditions with respect to some standard. 3) An evaluation of established procedures.

product—A product is the result of a process. A product is normally thought to have physical, tangible properties (a mixer, a design report). A service may have intangible properties (feels better, looks right).

product audit—1) An audit of a product or service (*see* audit). 2) Activity such as measuring, examining, testing, or gauging one or more characteristics of a product or service, done by an independent organization and comparing the results with specified requirements. 3) An independent examination of the characteristics and attributes of a product or service against a specification or acceptance criteria. 4) A quantitative assessment of conformance to required product characteristics.

qualitative—Of, relating to, or involving quality or kind. For example, qualitative analysis determines kinds of chemicals in a substance.

quality—1) Degree to which a set of inherent characteristics fulfills requirements. 2) Conformance to requirements. 3) Meeting customer requirements or achieving customer satisfaction. 4) Quality for the supplier is getting it right the first time and quality for the customer is getting what he was expecting.

quality assurance—1) The part of quality management focused on providing confidence that quality requirements will be fulfilled. 2) All the planned and systematic activities implemented within the quality system and demonstrated, as needed, to provide adequate confidence that an entity will fulfill requirements for quality.

quality audit—Systematic and independent examination to determine whether quality activities and related results comply with planned arrangements and whether these arrangements are implemented effectively and are suitable to achieve objectives.

quality control—1) Techniques and activities, such as inspection, used to verify conformance to requirements. 2) The part of quality management that focuses on fulfilling quality requirements. 3) Operational techniques and activities that are used to fulfill requirements for quality.

quality management—1) Coordinated activities to direct and control an organization with regard to quality. 2) Includes all activities of the overall management function (management system) that determine the quality policy, objectives, and responsibilities, and their implementation.

quantitative—1) Of, relating to, or expressible in terms of quantity. 2) Of, relating to, or involving the measurement of quantity or amount. For example: quantitative analysis determines the amounts of chemicals in a substance.

record—1) Data generated as a result of an activity or process. A record can verify that the activity took place. 2) A document stating results achieved or providing evidence of activities performed.

reliability—Lack of unplanned failures or shutdowns; that which one can depend on.

requirement—Need or expectation that is stated, generally implied, or obligatory.

root cause—The most basic reason for the effect, which if eliminated or corrected would prevent the effect from existing or occurring.

service—1) A process. 2) A value-added activity (value to the customer). 3) Intangible product that is the result of at least one activity performed at the interface between the supplier and the customer. 4) The occupation or function of serving. 5) Contribution to the welfare of.

shall—The word "shall" is used in requirement or contractual standards to indicate an absolute or strict requirement. The words "must" and "will" are also used to indicate an absolute or strict requirement.

standard—1) Something established by authority, custom, or general consent as a model (example: criterion). 2) Something set up and established by authority as a rule or the measure of quality, weight, extent, or value.[6] Note: the word "standard" is very general and includes documents such as procedures and specifications. It is also interesting to note that the use of the word "standard" as a noun has 18 different dictionary definitions.

suitable—Appropriate from the viewpoint of propriety, convenience, or fitness. 2) Right or appropriate for a particular person, purpose, or situation (ISO/TC 176/SC1 N274).

supplier—Organization or person that provides a product or result of a process. For example: retailer, distributor, manufacturer, or service provider.

system—1) A group of processes supported by an infrastructure to manage and coordinate its function. 2) A set of interrelated or interacting elements.

system audit—An audit of a system. Sometimes called a quality audit or environmental audit.

team—Two or more people working together to achieve a desired goal.

top management—Person or group of people who directs and controls an organization at the highest level. Synonyms are: executive, senior management, company officer, partner.

tracing—Audit tracing is following the chronological progress of a process. It is an effective means of collecting objective evidence. Forward tracing starts at the beginning; reverse (or backward) tracing starts at the end and works toward the beginning.

validation—1) Confirmation that a product or service will perform as expected or specified (for example pump performance test, vehicle road testing, tryout of software features). 2) Confirmation, through the provision of objective evidence, that the requirements for a specific intended use or application have been fulfilled.

verification—1) Confirmation, through the provision of objective evidence, that specified requirements have been fulfilled. 2) The act or process of verifying or the state of being verified; the authentication of truth or accuracy by such means as facts, statements, citations, measurements, or attendant circumstances.

work environment— A set of conditions under which work is performed. For example: temperature, lighting, pressure, humidity, space, psychological stress, and so on.

work instructions—A document that provides detailed information for carrying out a process, subprocess, or activity in a step-by-step manner (the document can be in any medium).

working papers—Documents, forms, checklists, or guidelines used by the auditor to help him/her perform an effective audit.

References

American Heritage Dictionary, 2nd ed. (Boston: Houghton Mifflin, 1985).

ANSI/ISO/ASQC Q10011:1994. *Guidelines for Auditing Quality Systems* (Milwaukee: ASQ Quality Press, 1994).

ASQ/ANSI/ISOC A8402-1994. *Quality Management and Quality Assurance–Vocabulary* (Milwaukee: ASQ Quality Press, 1994).

ASQ/ANSI/ISO Q9000:2005. *Quality management systems–Fundamentals and vocabulary* (Milwaukee: ASQ Quality Press, 2005).

Besterfield, D.H. *Quality Control*, 5th ed. (Columbus, OH: Prentice-Hall, 1998).

Hutton, D. *From Baldrige to the Bottom Line* (Milwaukee: ASQ Quality Press, 2000).

ISO 14001. *Environmental management systems–Requirements with guidance for use* (Geneva: International Organization for Standardization, 2001).

ISO 19011. *Guidelines for quality and/or environmental management systems auditing* (Geneva: International Organization for Standardization, 2001).

ISO/TS 16949:2002. *Quality management systems automotive suppliers.*

Juran, J.M. *Juran's Quality Control Handbook*, 4th ed. (New York: McGraw- Hill, 1988).

Mills, C.A. *The Quality Audit* (New York: McGraw-Hill, 1989).

Okes, Duke and R. T. Westcott, eds. *The Certified Quality Manager Handbook*, 2nd. ed. (Milwaukee: ASQ Quality Press, 2001).

Parsowith, B.S. *Fundamentals of Quality Auditing* (Milwaukee: ASQC Quality Press, 1995).

Random House College Dictionary (New York: Random House, 1988).

Russell, J.P. *Continual Improvement Assessment Guide: Promoting and Sustaining Business Results* (Milwaukee: ASQ Quality Press, 2004).

Russell, J.P. and T. Regel. *After the Quality Audit: Closing the Loop on the Audit Process*, 2nd ed. (Milwaukee: ASQ Quality Press, 2000).

Russell, J.P., ed. *The ASQ Auditing Handbook*, 3rd ed. (Milwaukee: ASQ Quality Press, 2005).

Russell, Paul and Rachel. *Business IQ* (Orlando: Fina Press, 2024).

Tague, Nancy. *The Quality Toolbox, Third Edition* (Milwaukee: Quality Press, 2023).

Webster's Third New International Dictionary, Unabridged (Springfield, MA: Merriam-Webster, 2002). http://unabridged.merriam-webs com (Feb 1, 2007).